A Robot
In Every Home

A Robot
In Every Home

An Introduction to Personal Robots
& Brand-Name Buyer's Guide

by Mike Higgins

KENSINGTON
PUBLISHING COMPANY
OAKLAND, CALIFORNIA

Published by
Kensington Publishing Company
6300 Telegraph Avenue
Oakland, California 94609
U.S.A.

Copyright © 1985 by Mike Higgins

Library of Congress Cataloging in Publication Data

Higgins, Mike, 1950-
    A robot in every home.

    Bibliography: p.
    Includes index.
    1. Robots. 2. Robots—Purchasing. I. Title.
II. Title: Personal Robots.
TJ211.H5   1985   643'.6   84-28915
ISBN 0-931445-15-9
ISBN 0-931445-16-7 (pbk.)

First Edition
987654321
Printed in the United States of America

To my mother, Betty, who taught me to read; and to my wife, Marlene, who encouraged me to become a writer.

"If every instrument could accomplish its own work, obeying or anticipating the will of others . . . if the shuttle could weave and the pick touch the lyre without a hand to guide them, chief workmen would not need servants, nor masters slaves."

Aristotle, Politics Book I

Contents

Acknowledgments

I would like to thank everyone who helped me in the research, writing, editing, printing and publication of this book, including:

Jonnie Bryant, Raymond Côté, Bob Cremer, Mike Haughey, Jan Kessner, Jo Liana King, Jackie Klein, Vic Kley, Ray Lindsey, Laurie & Page Milliken, Kent Myers, Marilyn Oskard, John Peers, Lory Poulson, Richard Prather, Tim Sakamoto, Bruce Sanchez, T. Gray Shaw, Sharon Smith, Dan Talken, Walt Tunick, Dennis Van Dusen, Wayne Wilson, and Mary Woodworth-Haglund.

Special thanks to Linda Byrne, of Steve Rank Inc., for designing the cover, and to Steve Rank for giving me advice and encouragement throughout this entire project.

Part I

Personal Robots: Past, Present and Future

A   al robot is easily the most fascinating type of robot,
and t   st versatile—capable of performing many of the
functi   the other types we have discussed. It's a machine
that se   have a life of its own, running around your
house,   to you and performing various services for you
at your   nd. There are now several dozen manufacturers
of perso   ots, robotic components and software, and
robot sho   beginning to spring up in major cities across
the countr

Once pe   et over the surprise of finding out that the first
home robo   already available, they usually have another
surprise wai   r them. Most people vastly overestimate the
capabilities   first generation of personal robots. We've
been conditi   by so many years of Hollywood robots
(most notably   bie the Robot from *Forbidden Planet* and
C3PO from *S   Vars*) that we are disappointed when we
don't see that   of magic in the real world. Before you
decide to buy a   onal robot, you need to know what's
possible and wha   t.

Joseph Engelbe   who founded the first industrial robot
manufacturing co   v over 20 years ago, has given us an
elegant analogy of   nitations of today's robot technology.
To get a first-hand i   sion of the difficulties involved, says
Engelberger, start by   ring a thick layer of petroleum jelly
on the lenses of a pai   ectacles. Then put the spectacles on,
tie one hand behind y   ck, put a mitten on the other hand,
pick up a pair of cho   s, and try to assemble something
according to a detailed   instructions!

What *can* personal r   lo, then? At the most basic level,
they can move, sense th   ironment, and manipulate that
environment. The most   ting ones can also speak and
understand English. Wit   opriate software, these basic
functions can be put toget   n infinite variety of combinations, enabling a personal r   entertain you, educate you,
and act as a companion, se   uard and servant. The key
phrase here is "with appropi   ftware." Without adequate
applications software, a robo   a computer on wheels.

The Future of Personal Robots

In many respects, the personal *robot* industry is about where the personal *computer* industry was 10 years ago. Today's personal robots are not the perfect mechanical servants we would like them to be, but we have to start somewhere. When the Wright Brothers flew the world's first airplane in 1903, who could have foreseen the development of the Boeing 747 less than 70 years later?

Actually, we have an advantage over the people who struggled to popularize early flying machines, telephones, automobiles and television receivers. The Wright brothers didn't know they were building the precursor to the 747, but we have a pretty good idea of where the personal robot industry is headed. People want mechanical servants to perform all the tedious jobs around the home that nobody likes to do. It may take us 20, 50 or 100 years, but we will eventually succeed in mass-producing fully functional mechanical servants.

How to Use This Book

This book is divided into two sections. Part I describes how personal robots were developed, what's inside them, and what they can be used for in your home. Part II is a brand-name guide to the specific characteristics and capabilities of each of the major brands of personal robots currently available.

At the back of the book, you'll find lists of manufacturers, retailers, associations and other resources connected with the personal robotics industry. You'll also find an explanation of the technical terms used in the robot industry, a list of books on the subject of personal robots, and an index. Now, let's begin with a look at how personal robots came into existence.

2

The History of Robots

The development of functioning, intelligent robots is the culmination of thousands of years of human endeavor. Robots could not have existed without the level of sophistication achieved only in the last hundred years in mechanical engineering and in the last ten years in computer science.

The history of computers starts with the abacus, which was used more than 5,000 years ago. The oldest mechanical artifact found to date is a model of a Saluki hunting dog, discovered in an Egyptian tomb dating back to about 2000 B.C. The dog's mouth opens when a lever in its stomach is operated.

From these simple beginnings, the technologies of mechanical engineering and computer science have advanced to the point at which a self-propelled, self-directed mechanical device can move around your home, hearing, understanding and obeying your spoken commands.

Early Automatons

Automatons—decorative or amusing mechanical devices that run unattended once they are turned on—were the predecessors of today's personal robots. One of the earliest known

23

automatons was a wooden model of a pigeon said to have been made in 350 B.C. by Archytas of Tarentum, a friend of Plato. Steam or compressed air was apparently used to make the bird revolve on a pivoted bar.

One of the first devices designed to perform a useful function rather than just to amuse people was a mechanism for opening and closing the doors of a temple, invented by Hero of Alexandria around 100 A.D. The device relied on fire to cause air to expand, forcing water into a bucket. The extra weight of the water caused the bucket to descend, pulling ropes that turned rollers and opened the doors. When the fire was extinguished, the air would contract, siphoning the water back out of the bucket. A counterweight caused the doors to close again.

Water was also the source of power for a clock built at Gaza, Syria, around 500 A.D. The clock featured a figure of Hercules using a club to strike the hours. Such was the interest in water-driven automatons in China in the sixth and seventh centuries that a book was published on the subject, with a title that translates as "Book of Hydraulic Elegancies."

In Mesopotamia, in the thirteenth century, al-Jazari designed model peacocks operated by water. The same man is credited with writing a book describing an automaton that could automatically fill and empty a wash basin. After the user washed his hands, he would pull a lever, causing the dirty water to be drained. It would be replaced by fresh water poured into the basin by the figure of a woman.

Two of the most famous examples of automatons are the parade of puppets made for the clock of the Frauenkirche at Nuremberg in Germany, completed in 1361, and the two bronze figures that strike the hours in the clock in Venice's Piazza di San Marco, built in 1499.

In the sixteenth century, elaborate automatons incorporating extravagant special effects became highly fashionable among wealthy Europeans. One example was a 16 foot high clock sent to the Turkish Sultan by Queen Elizabeth I of England in 1599. The clock included bells, an organ, mechanical trumpet-players, and blackbirds that sang and moved about.

Al-Jazari's wash-basin automaton.

The development of spring-driven clockwork mechanisms in the eighteenth century led to the creation of some incredibly complex automatons. In 1738, the Frenchman Jacques Vaucanson demonstrated a duck that could walk, swim, flap its wings, peck corn and even "digest," or at least dissolve, the corn it ate. He also exhibited a flute player with a mechanism for manipulating the fingers to actually play the flute using air blown from a bellows through the players' mouth. Each figure was life-sized and mounted on a pedestal containing the mechanism that drove it.

Among Vaucanson's followers were the Swiss Pierre Jacquet-Droz and his son, Henri-Louis. One of their most famous figures was of a woman using all ten fingers to play music. Another figure was reportedly capable of writing any message up to 40 characters long.

Jacquet-Droz writer automaton.

MUSÉE D'ART ET D'HISTOIRE, NEUCHATEL, SWITZERLAND

Jacquet-Droz music-player automaton.

Then, in 1769, came the famous Turk chess player hoax. Baron Wolfgang von Kempelen exhibited an "automatic" chess player that was, in fact, controlled by a man hidden inside the machine. Many people were fooled by the device, which contained a very elaborate mechanism that did not actually do anything.

As the nineteenth century began, the emphasis shifted from clever toys to automatic machinery that could perform useful work. In 1804, Joseph-Marie Jacquard invented a method for controlling an industrial loom using cards with holes punched in them. This was one of the most significant inventions of the Industrial Revolution. The entire computer industry was to be based on the punched card, but of course no one had any way of knowing that at the time. It's worth noting, in passing, that Mary Shelley's "Frankenstein" was published in 1818—the first of many stories to be written on the creation-turns-on-creator theme.

In 1894, Thomas Edison built a miniature version of his phonograph into a small doll. The demand was said to be so great that his factory was turning out 500 a day and still barely keeping up with sales. By the beginning of the twentieth century, all the elements of a personal robot were generally available, with the exception of some kind of built-in controlling intelligence. So let us now turn our attention to the development of the "mechanical brain," as computers were called when they first came out.

✱The Development of the Computer

For thousands of years, the abacus remained the only tool designed to augment our ten fingers as calculating aids. The first step toward automating the tedious process of performing complex calculations by hand was taken in the seventeenth century, when John Napier invented logarithms. The time-consuming and error-prone process of multiplying or dividing large numbers by hand could now be replaced by the much more efficient process of looking up the logarithms of the numbers in a book of tables and simply adding (or, for division, subtracting) the logarithms.

Thomas Edison's phonograph doll.

In 1633, William Oughtred, an English mathematician, hit upon the idea of printing numbers next to their logarithms on a piece of wood. By placing two such pieces of wood next to each other, the numbers and their logarithms could be manipulated without the need for books of tables. Thus was born the slide rule, indispensable equipment for every engineer until the electronic pocket calculator became widely available in the 1960s.

But we are getting a little ahead of our story. The first mechanical calculator was built by Blaise Pascal in 1642 to perform calculations for his father's business accounts. The machine used wheels with the digits 0 through 9 printed on them. When a wheel was turned past the "9" position, it would engage a cog to advance the position of the next wheel to its left. This is, of course, the same process we use when adding numbers together using pencil and paper: "9+6=15; write down the 5 and carry the 1." Mechanical desk calculators very similar to Pascal's were used in businesses everywhere until they were superseded by the same invention that killed the slide rule.

In 1694, Gottfried Willhelm Leibniz completed his "Stepped Reckoner," a machine that could multiply, divide and extract square roots as well as add and subtract.

In 1822, Charles Babbage devised his "Difference Engine" and applied to the British government for funds to develop it. He received thousands of pounds—a considerable sum of money in those days—but was unable to perfect his invention. Later, he devised a more advanced machine he called the "Analytical Engine."

The Analytical Engine was more than a hundred years ahead of its time. The design embodied all the major functional components of every electronic digital computer in use today: programs to tell the machine what to do; data on which those functions were performed; sequential control with conditional branching and looping; arithmetic and storage units; input and output mechanisms; and the ability to modify its own programs and data. Unfortunately for Babbage, his theoretical design was too far ahead of the mechanical technology available in his day, and the machine was never built.

In 1854, George Boole invented the algebra of logic, now called Boolean logic, which showed how problems in logic could be solved mathematically. Boolean logic uses only the symbols "0" and "1," and forms the operational basis for all modern digital computers.

Then, in 1886, Herman Hollerith adapted Jacquard's punched card concept to tabulate the 1880 U.S. census. His methods were so efficient that the time taken to complete the census was reduced by two-thirds. In 1911, Hollerith cofounded the Computing Tabulating Recording Company, which later became International Business Machines—now known universally as IBM.

In 1939, Howard Aiken of Harvard University began working with IBM engineers on the Automatic Sequence Controlled Calculator, also called the Harvard Mark I. This was a five-ton monster of a device, controlled by punched paper tape and relying on over 3,000 electromechanical relays for its memory.

The first purely electronic computer was the ENIAC (for Electronic Numerical Integrator and Calculator), which was invented by Presper Eckert and John W. Mauchly, and completed in 1947. The ENIAC consisted of enough machinery to fill a large room, and used thousands of vacuum tubes as its memory. It was thousands of times faster than electromechanical calculators and could, in a few hours, perform calculations that would otherwise have taken years to do. Although it was much faster than any calculating machine built before it, the unreliability of all those vacuum tubes resulted in frequent breakdowns.

In 1947, the transistor was perfected at Bell Labs by John Bardeen, Walter Brattain and William Shockley, who were awarded the Nobel prize in recognition of their achievement. This was the first step in that cycle peculiar to the computer industry, in which components continually become orders of magnitude smaller and faster, while simultaneously dropping in price and becoming more reliable.

The world's first commercially produced computer was the UNIVAC 1 (for Universal Automatic Calculator), of which 45 were manufactured and sold to the U.S. Census Bureau and to General Electric Company. In 1952, IBM added its first computer to the company's line of tabulating machines. By the 1960s, IBM had become the leading supplier of computers, introducing the highly successful IBM 360 in 1964.

As electronics manufacturing techniques improved, it became possible to cram hundreds of transistors onto a single "chip" of silicon approximately a quarter of an inch square. This resulted in a new generation of computers called "minicomputers," which were more compact than the "mainframe" computers that came before them. In 1969, Intel Corporation designed a single chip, called the 4004, that could perform some simple control functions. The processing capacity of the chip was subsequently doubled, and the new chip was dubbed the 8008. In 1974, Intel upgraded the 8008, making it into the world's first microprocessor—the 8080.

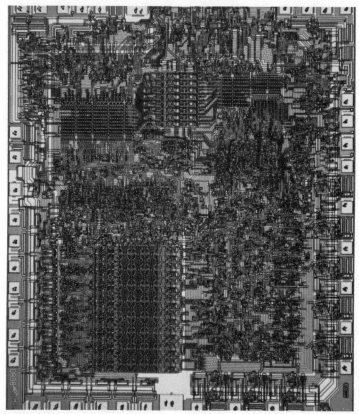

Intel 8080 microprocessor.

The microprocessor was written up in all the computer and electronics magazines as an engineering marvel. Everyone was impressed by the accomplishment, but no one knew quite what to do with it. Then, in 1975, *Popular Electronics* magazine ran an article on how to build a computer from a kit produced by a small electronics development company in Albuquerque,

New Mexico. The company was called MITS, the computer was the Altair, and thousands of orders poured in from readers of the article. This was the birth of the personal computer industry. MITS was later sold and, ironically, both its name and that of the Altair have disappeared completely except for the history books.

At first, anyone interested in personal computers had to be something of an electronics engineer and computer design expert. All this changed in 1976, when Steve Wozniak and Steve Jobs started Apple Corporation using the money raised from the sale of Jobs' VW van. The company's Apple II computer made it possible for a non-expert to buy a computer, take it home, and immediately start running programs on it. The age of the personal computer had arrived.

One of the early Apple computers.

The Rise of the Robot

The word "robot" was first used in a play by Karel Capek called *R.U.R.* (for "Rossum's Universal Robots"), produced in 1921 in Prague, Czechoslovakia. Capek derived the word from the Czech *robota*, meaning "forced labor." In the play, man-like machines are invented to relieve men of the need to perform manual labor, but the machines turn against their masters and kill them. The play was produced in New York in 1922, where it was successful in spite of poor reviews, mainly because of the power of its central theme.

Robots turn on their masters in Capek's 1921 play, R.U.R.

In 1941, in his science-fiction story "Runaround," Isaac Asimov coined the word "robotics" and formulated his now-famous Three Laws of Robotics, which read as follows:

1. A robot may not injure a human being, or, through inaction, allow a human being to come to harm.

2. A robot must obey the orders given it by human beings except where such orders would conflict with the First Law.

3. A robot must protect its own existence as long as such protection does not conflict with the First or Second Law.

Asimov used the interaction between the simple, straightforward logic of these rules and the complex, often illogical, real world as the basis for more than a dozen short stories and three full-length novels about robots. The thought-provoking, often highly philosophical themes of Asimov's stories set them apart from the tradition of mindless mechanical-monster-out-of-control plots created by the Hollywood movie industry.

Asimov never dreamed that robots would make the transition from fantasy to reality within his lifetime, but they have, of course, done exactly that.

The world's first robot was not something you would recognize from seeing Capek's play or reading Asimov's science-fiction stories. It was an industrial robot, invented in 1954 by George Devol, which looked much more like a huge mechanical version of a man's arm than like a mechanical man. Industrial robots evolved from numerical-control machines, which are machine tools that can be programmed, using punched tape, to perform a specified sequence of movements automatically.

In the 1960s, an experimental robot called Shakey was designed and built at Stanford Research Institute (now known as SRI International). With its three-wheeled base, consisting of two drive wheels and a caster, radio link to an external

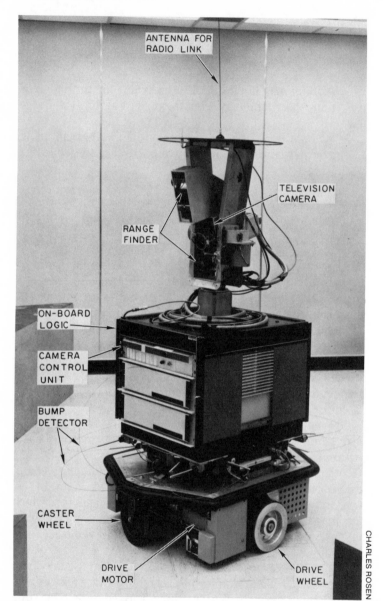

Shakey experimental robot.

computer, "bumper" sensors, automatic rangefinder and swivelling head containing a TV camera, Shakey was the precursor of today's personal robots. Shakey could perform a number of experimental tasks, such as learning the boundaries of a room and locating objects inside the room.

Just as in the early days of the personal computer, the first people to build personal robots were the hobbyists. Working alone or in small groups with other hobbyists, they built their own working, mobile robots from surplus and scrap materials, with inexpensive personal computers serving as the robot's "brain." They did it for the fun of it, with no thought of commercial application or profit.

The world's first commercially available personal robot was the RB Robot Corporation's RB5X, which was introduced to the general public in September 1982 and available in some stores by November that same year. Heath Company introduced its Hero 1 the following month, and in January 1983, Androbot Inc. unveiled its Topo home robot. The age of the personal robot had arrived.

3

Inside the
Personal Robot

What does a personal robot consist of? There is no single answer to this question that everyone would agree with. What we can do, though, is look at some of the attributes of a personal robot that most people would at least consider including in a definition.

Let's start with the ability to move around. Although most industrial robots are mounted on an immovable platform, all personal robots are mobile (with the exception of miniature versions of the industrial robots used in educational applications). And when a personal robot has moved itself to a particular location where some task must be performed, it will also need an arm or a hand of some kind with which to perform that task.

Of course, a robot won't be going anywhere unless it has a source of energy to power its movements and a brain to control them. The robot bases its decisions on rules of behavior (the robot's programming), its experience (the results, stored in memory, of previous decisions) and information about the robot's environment (sensory input). Robots can duplicate to some extent all the familiar human senses of sight, hearing, touch, smell and taste, and can even improve on nature by sensing physical quantities that we are unable to detect.

Mobility

The design of a robot's motion system is critical for several reasons. If the robot can't travel to wherever a job needs to be done, it certainly won't be able to do that job. And, since moving the robot takes more energy than any other robotic function, the motion system must be as efficient as possible. In addition, the motion system must not cause any damage to the surfaces the robot moves on (e.g., by pulling threads out of your rugs and carpets, or leaving black marks on your tile and linoleum floors).

It turns out to be a big mistake to try to emulate nature by designing a robotic motion system that relies on legs, especially a system that uses only two legs. Walking on two legs is actually a kind of controlled falling—just before you fall over, you move one leg in front of the other to save yourself. With four or more legs, you can move one or more legs at a time and still leave three legs on the ground for completely stable support. But multi-legged motion systems are so complex and costly that they have been used only in a few experimental robots built with government research funds.

Several fictional robots have been equipped with a scaled down version of the tracks used on bulldozers, tanks and other vehicles designed for rugged terrain. The concept is appealing, but impractical because of the damage tracks cause to the surface they run on—especially during a turn.

This leaves us with wheels, which are safer and simpler, and which use less power than systems based on legs or tracks. All commercially available personal robots use wheels, although there are still some design compromises to be made. With small wheels, you get smooth motion and very little strain on the drive mechanism. The problem comes when the robot encounters bumps and uneven surfaces. With bigger wheels, you get better performance on rough terrain, but the robot will require a more powerful drive system and may not start and stop as smoothly.

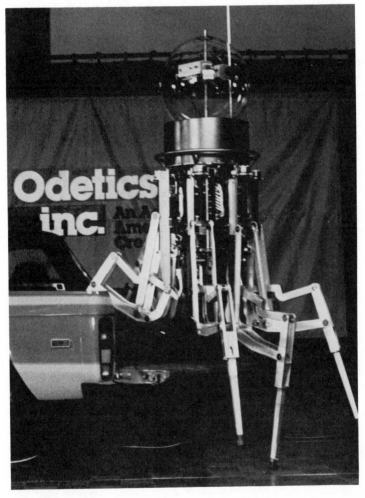

Six-legged robot.

ODETICS INC.

The most stable configuration uses three wheels equally spaced. Using this configuration, there are several different ways to drive and steer the robot. One way is to drive the two rear wheels independently, leaving the single wheel at the front undriven and able to pivot freely. A drawback of this method is erratic steering if one of the driven wheels should skip on a slippery surface.

Another approach is to drive only the front wheel. This design is economical and it allows the robot to make very tight turns, but the front wheel must be able to carry its share of the weight of the robot, as well as that of the drive motor, while still being able to pivot.

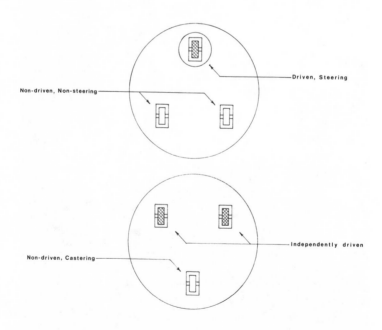

Drive systems based on three wheels.

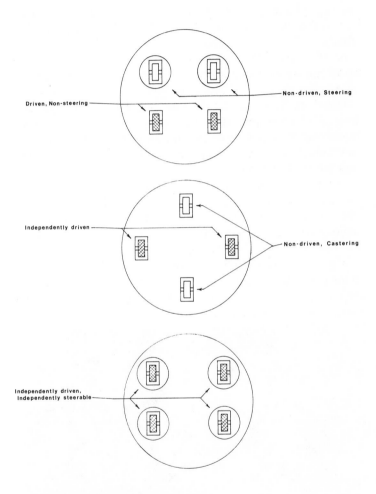

Drive systems based on four wheels.

By adding a fourth wheel, you get improved stability and more ways of steering and driving each wheel. Some designs require differential gearing on the drive mechanism (so that a driven wheel on the inside of a turn can rotate more slowly than the wheel on the outside of the turn) and compensated steering linkages (so that an inside steering wheel can make a turn of smaller radius than that made by the outside steering wheel).

Another possibility is a four-wheeled system in which each wheel can be independently steered and driven. This kind of system needs a lot of control electronics, but it is very versatile.

These are by no means the only possible wheel configurations. Androbot came up with a two-wheeled system for the Topo robot, in which the wheels are angled at 30 degrees to the horizontal. Two rubber bumpers prevent the robot from tipping too far forward or back.

Other esoteric drive systems rely on wheels within wheels, such as the Metamotion system invented by Bill La. The system consists of three independently driven wheels with rims made of rollers, which enable the whole device to move in any direction and at any orientation.

A university research team even came up with a one-legged robot that looks like an automated pogo stick. At the end of each hop, the robot can sense which way it is tipping and correct for it in the next hop.

METAMOTION LABORATORIES

Wheels-within-wheels drive system.

Arms and Hands

Robotic arms, whether for industrial robots or for personal robots, tend to emulate the human arm. A typical design includes a shoulder joint, upper arm segment, elbow joint, lower arm section, wrist and hand (usually in the form of a simple two- or three-finger gripper).

Robotic arms can do things human arms could never do, such as rotating through more than 360 degrees, but they move much more slowly. No robotic hand yet developed comes close to having the sensitivity and manipulative ability of the human hand, but robots have the advantage of being able to function with any one of a large number of interchangeable hands.

Not all personal robots available today are equipped with arms, and even when an arm is available, it's often an optional accessory. This is because arms are relatively expensive to manufacture and currently do not perform as well as most people would want them to. When the arm is priced separately, you can decide for yourself whether it's worth paying the extra money for it.

Even robots that do not have functioning arms usually have at least some provision for allowing the robot to carry small items from one place to another in your home.

Energy Source

The power for driving a personal robot's wheels and moving its arms is invariably provided by a battery. Industrial robots can use hydraulic and pneumatic power sources, but these alternatives are out of the question for a small, mobile robot. Deriving energy from combustion is similarly unrealistic and, in any case, these non-electrical energy sources would be useless for powering the robot's electronic equipment. Solar power would require much too large a surface area, and trailing a power cord plugged into an electrical outlet is extremely inconvenient.

The amount of power needed to move an object increases with the weight of the object and the speed at which you want to move it. With existing battery technology, you can expect to run a typical personal robot for about three to four hours after recharging its batteries overnight.

One of the most reliable, relatively inexpensive batteries available is the lead-acid battery universally used as a car battery. The trouble with this type of battery, however, is that it's heavy, it must remain upright, and it can leak acid as well as emitting explosive gases. You have to add water from time to time, and even the so-called "maintenance-free" batteries are not entirely immune to these problems.

Nickel-cadmium batteries are truly maintenance-free, but they are expensive and sensitive to temperature. They also suffer from a "memory" problem: if you repeatedly recharge them before they really need it, they will "remember" this charge level and refuse to deliver power beyond that level in the future.

Rechargeable alkaline batteries are also maintenance-free, but they cannot be recharged as easily or as often as other types and they provide less power for their weight.

The most useful type of battery is the sealed, maintenance free, gelled electrolyte lead-acid battery. It can be used and recharged in any position, is not overly sensitive to temperature, does not suffer from the "memory" effect, can be recharged hundreds of times, and provides a lot of power for its weight.

The Robot Brain

A personal robot's brain is a small computer consisting of a central processing unit, some memory, and a means of communicating with the outside world. Like any computer, the robot brain has the capacity for receiving information, storing it, deciding what to do next, and transmitting information as a result of that decision. Before the robot can do any of this, however, it must be programmed.

The information received, processed and transmitted by an ordinary computer is usually in the form of characters entered from a keyboard, manipulated in some way, then sent back out to a disk drive, display screen, or printer. A robot is different from a computer in that it receives its information from sensors that measure various aspects of the robot's environment. The robot processes this information, makes a decision, and transmits instructions to its motors and other subsystems. As a result of this sequence, the robot's environment changes. Then the whole cycle is repeated. This process is called feedback, and it's a very important concept.

A familiar example of a feedback process is the thermostat. This device measures the temperature in a room, makes a decision as to whether the room is colder than it's supposed to be, and turns on the heater if necessary. The thermostat continues to monitor the room's tempererature, and will turn the heater off again when the room becomes warm enough.

Personal robots work in a similar fashion. For example, many robots are programmed to monitor their battery charge level. If the level drops below a certain value, the robot may respond by shutting down non-essential functions to conserve energy until it detects that the battery has been recharged to a safe level.

Sensors

Just as the room thermostat needs a way of measuring the temperature before it can do its job, personal robots need various sensors so that they can make sensible decisions about what to do next. Although personal robots have not been around for very long, the standard senses of sight, hearing, touch, taste and smell can all be duplicated to some extent, and other senses not found in nature can also be added.

Vision

Vision, the most important of the human senses, is also the most desirable sense for a robot to have. Providing robots with vision is a very difficult design challenge, however, although there have been some recent technological improvements. Two-dimensional vision is now quite commonplace in industrial robotic applications, but achieving and any kind of sophisticated vision for personal robots continues to present considerable problems.

A robotic eye based on a video camera is not the simple

solution you might expect. It's easy enough to mount a camera on top of a robot, but the hardware and software needed to process the image formed by the camera are very complicated and expensive.

The first task in image processing is to digitize the image. This is not too difficult—all you have to do is split the image up into an array of cells (256 by 256 is typical) and assign a number to each cell corresponding to the amount of light falling on that cell. The next task is to search the digitized image for lines representing the edges of an object, which is a little more difficult. Many clever mathematical schemes have been developed for identifying edges by discerning sudden changes in brightness between adjacent cells.

The final step is to take the shape formed by all the edges and compare it with shapes stored in memory. If a match is found, the robot has successfully identified the object it is looking at and can use that information in deciding what its next action should be.

This process is difficult enough when only two dimensions are considered, such as when an industrial robot is looking down at a conveyor belt carrying machine parts. It's relatively easy to identify the shape of a nut or bolt under controlled conditions. But consider the effect of a hexagonal nut going by on the conveyor belt in an upright position, instead of lying flat. The robot would see a rectangle (the top of the nut) instead of a hexagon with a hole in the middle.

This problem of three-dimensional to two-dimensional transformation would obviously be much worse in the uncontrolled environment inside your house. Shadows and patterns of reflectivity that shift during the course of the day also contribute to the difficulty of implementing three-dimensional vision.

Another problem is the amount of data that must be stored and processed. If you use one byte of memory to store a number corresponding to the amount of light falling onto each cell in a 256×256 array, you'll need 65,536 bytes to store a single image. It makes sense to store one image for processing while the next image is being assembled, so make that 131,072 bytes for image storage. That's quite a lot of memory.

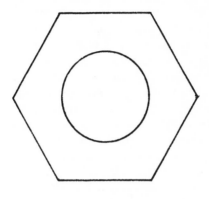

TOP VIEW

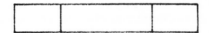

SIDE VIEW

Two very different views of a hexagonal nut.

If you wanted to process 30 images per second, the rate at which a TV set presents individual "frames" to give the impression of continuous motion, you would be processing close to 2 million bytes of information every second—a very high processing speed even for today's most sophisticated computers. Fortunately, the computer industry continues to increase computer speed and capacity each year while simultaneously reducing size and cost, so this kind of computing power should soon be readily available at an affordable price.

In the meantime, simpler forms of vision are available for personal robots. A photoelectric cell, which generates a current that varies with the amount light falling on it, can be used to direct a robot toward a light source. The robot need only be programmed to turn and find the position at which the cell generates the most current. A homing function could be implemented by placing the light source at the robot's home position and programming the robot to move toward the light.

A very inexpensive vision system can be implemented using only a photodetector and some reflective tape. You simply lay the tape on the floor to lead the robot along the path you want it to take. The robot can easily detect whether or not its photodetector is over the tape, and move sideways to reposition itself as necessary.

Touch

Almost every personal robot has touch sensors. The most common type is a simple "bumper" detector—a microswitch connected to a bumper pad or wire feeler, similar to a cat's whisker. When the robot bumps into an object, the microswitch closes and the robot gets the message that bumper number so-and-so has touched something. The robot's usual response will be to abandon further attempts to move in that direction and to try another route. If touch sensors are located on the robot's grippers, they can serve to tell the robot that it has grasped an object and can proceed to lift it.

A more sophisticated touch sensor would also sense the force of the contact with a pressure transducer, a device that converts physical pressure into electric current. Some university researchers are working on a touch-sensitive robotic skin made of a special film that acts like an array of pressure transducers. The shape of an object can be determined by using a microprocessor to analyze the signals produced when the object is pressed against the skin. Touch sensors are not very glamorous, but they are simple, effective, and inexpensive enough to be installed in many locations around a robot's body.

3 *Smell* Protect people

The easiest way to equip a robot with a sense of smell is to use an adaptation of a device already familiar in most households: the smoke detector. These detectors "smell" smoke by counting the number of charged particles in the air. When the number rises, this is an indication that more particles are present—particles that constitute smoke. The ability to detect smoke in the air and alert the occupants of a house to the danger of fire would certainly be useful in a personal robot.

Gas detectors can also be built into robots, to detect the presence of noxious fumes. Some robot manufacturers have talked about equipping robots with ammonia "sniffers" so that they can recognize when people are present by detecting the characteristic odor given off by the human body.

7 *Taste* Protect itself.

The sense of taste serves primarily to attract an organism toward appropriate food and away from nutritionally worthless food. Since robots don't get their energy from eating food the way people and animals do, they don't need a chemically-based sense of taste. But the robotic equivalent of a sense of taste could, perhaps, be implemented simply by adding a probe that the robot could use to check the voltage of an electrical outlet before using it to recharge its battery.

This would, at least, prevent the robot from remaining uselessly at a non-functioning outlet or damaging itself by plugging into an outlet of the wrong voltage.

Taste is not, obviously, one of the most important robotic senses.

1 *Hearing* UNDERSTAND commands

In its simplest form, a robotic sense of hearing can distinguish only between noise and no noise. It's an easy matter to add the ability to count the number of noises heard, and, for example, to perform a different task depending on the number of times you clap your hands. You could also simulate speech recognition by using commands such as "go" (one word), "now stop" (two words), "now turn left" (three words) and "turn to the right" (four words).

But the most interesting form of robotic hearing is a genuine ability to recognize and understand spoken commands. Speech recognition is accomplished by taking analog signals from a microphone, digitizing them, and comparing them with common words and phrases already stored in digitized form in memory. Recognition consists of successfully matching a spoken command to a stored word or phrase.

Speech recognition systems are classified as speaker-dependent or speaker-independent. With a speaker-dependent system, only one speaker can issue commands. If your robot is equipped with such a system, you first have to "train" it to recognize your voice. You repeat the command vocabulary several times so that the robot can create templates that reflect your particular voice characteristics. Then, when you issue a command, the robot can check it against the stored templates to find a match.

With a speaker-independent system, as the name implies, any speaker can be recognized. The manufacturer of such a system performs the training operation for you, using a composite derived from a wide range of different voices.

There are advantages and disadvantages to both types of systems. Although speaker-dependent systems are limited to only one user, this can be a benefit from the point of view of security. There's no chance that your robot will ignore your commands and obey a burglar instead. Also, with some systems, you can store several sets of templates and just tell the robot to use a new set when you want someone else in your family to control the robot.

The price of a speaker-independent system is usually paid in the form of a smaller vocabulary of recognizable words, or perhaps a lower success rate in recognizing words. Some speech recognition systems allow you to speak in a normal, conversational tone, whereas others require you to pause slightly between words. Because they must be able to understand a wide range of different speakers, speaker-independent systems are often limited to recognition of discrete words, requiring you to pause between words when talking to the robot.

All speech recognition systems have difficulty understanding people with poor enunciation, a strong accent, or even a bad cold. Loud background noise also makes recognition more difficult. For reasons of safety, a robot should be able to understand a "stop" command regardless of who issues it, and how panicky they sound when they say it. Another safety feature is to have the robot repeat a command back to you to make sure it understood correctly, and wait for your confirmation, before running off to polish the garbage or throw out your Ming vase.

Speech Synthesis

Although it's not really a sense in the way that sight, hearing, touch, taste and smell are senses, the ability to speak is one of the most useful and most appealing attributes of a personal robot.

For many years, robots have been given the appearance of being able to talk by the simple trick of equipping it with a radio receiver and paying a man to talk into a transmitter some distance away. A step up from this method is to install a tape recorder inside the robot and have the robot start and stop the tape under program control. Even though the robot does not have any control over what gets said, at least it can exercise some control over when.

The best way to give a robot the ability to speak, however, is through speech synthesis. A speech synthesis system consists of some memory for storing phrases, words, or portions of words in digital form, a digital-to-analog converter for converting the digital representations of the sounds into analog waveforms, an amplifier to increase the power of the signals, and a speaker to convert the electrical signals into sound.

Some systems start with complete words and phrases, whereas others are based on phonemes—the elementary sounds that make up all spoken language. For word-oriented systems, you can buy off-the-shelf chips containing specialized vocabularies for many different applications, or, with a phoneme-based system, you can generate an infinite variety of words or sounds from the basic building blocks.

Most personal robots use a phoneme-based system for greater flexibility. With either type of system, you make the robot talk by supplying the code numbers of the words or phonemes you want, accompanied by a "speak" command. Some manufacturers offer a text-to-speech software package that saves you the trouble of looking up codes for each individual word. A text-to-speech package will generate the correct phonemes for you within the constraints imposed by the idiosyncrasies of written English. (It will be a long time before a computer will be able to figure out the difference between the pronunciations of "bough" and "cough.")

Synthesized speech still lacks the quality of natural speech, but it is by no means limited to a flat monotone. Words and syllables can be modified by the addition of inflection and emphasis, resulting in much more realistic-sounding speech.

Extra-Sensory Perception

Robots can't detect the presence of ghosts or read people's minds (at least as far as we know!), but they do have the ability to sense some physical quantities that are beyond the limits of our senses.

Our eyes are sensitive to what is referred to as the visible portion of the spectrum, from red through orange, yellow, green, blue, indigo and violet. Robots can detect additional frequencies at both ends of this range: the so called infrared and ultraviolet frequencies. The infrared portion of the spectrum is particularly useful because it enables robots to see in what is, for us, darkness. Infrared light is emitted naturally from objects that are warmer than their surroundings, and it can also be supplied from an infrared source and detected as it is reflected from objects in front of the source.

Moving further away from the visual portion of the electromagnetic spectrum, we come to x-rays and radioactive emissions. Although x-ray vision would not be particularly useful in the home, the possibility exists of equipping a robot with this capability for use in hospitals or at airport security checkpoints. As for radioactivity sensors, let's all hope this is one kind of sensor we will never need on our home robots.

Ultrasonic transmitters and receivers are common on personal robots for use in detecting motion and measuring distances. These sensors, known as sonar sensors, work by bouncing high-frequency sound (much too high to be audible) off objects and measuring the time for an echo to be received. The speed of sound is so slow compared with the speed of a computer that the echo time, and therefore the distance, of an object can be determined with great accuracy.

Robots can make precise measurements of other physical quantities that we can only guess at, in addition to distances. Most of us have some sense of time, for example, although it does tend to be subjective. No robot can function without a computer to control it, and no computer can run without an internal clock for timing its internal operations. It's a simple matter to tap into that clock and use it for time-of day and calendar date functions, and many manufacturers have equipped their robots with the ability to wake you up in the morning or remind you of your wedding anniversary.

4

Robots as Entertainers
and Educators

It's easy to use a personal robot for running computer games or educational programs, because every personal robot already has a built-in microcomputer. The advantage of using a robot instead of a computer to run a game or educational program is that the robot can walk and talk. With these added abilities, robots make games more interesting and educational programs more realistic.

Robots as Entertainers

At the simplest level, a robot may be nothing more than an animated toy that can move around under its own power. The Tomy Dingbot is an example of this type of robot. It runs around until it bumps into something, then it stops, turns its head from side to side as if studying a map it's carrying, and moves off in a new direction.

With the addition of voice recognition, a robot will also obey your commands. The Tomy Verbot can be trained to recognize up to eight individual commands, such as go, back, right, left, stop, lift, lower and smile.

A Robot In Every Home

TOMY CORP.

The Dingbot.

TOMY CORP.

The Verbot.

More sophisticated toy robots can be taught to follow a certain sequence of movements, punctated by appropriate comments recorded in advance on a built-in audio cassette tape. The Tomy Omnibot can easily be programmed—by a child—to roll from your kitchen to your living room, stop in front of the couch with a tray in its hands and ask "Would anybody like a snack?"

The Omnibot.

The next step up is a robot preprogrammed at the factory with a repertoire of nursery rhymes, stories, jokes and interactive games. The Hero Jr. robot from Heath Company, for example, will ask you to think of an animal, then try to guess which one you picked by asking questions like "Does it have four legs?" and "Is it furry?" (which you answer by clapping your hands once for "yes" and twice for "no").

The Hero Jr.

When a robot has a complete personal computer inside it, including keyboard, monitor and disk drive, you can use it to play any game written for that particular type of computer, as well as any game written especially for the robot. The Hubotics Hubot comes with a built-in, fully CP/ M-compatible computer, and a software disk containing several word games, including the famous Eliza. (Eliza was written to demonstrate artificial intelligence techniques. The computer plays the role of

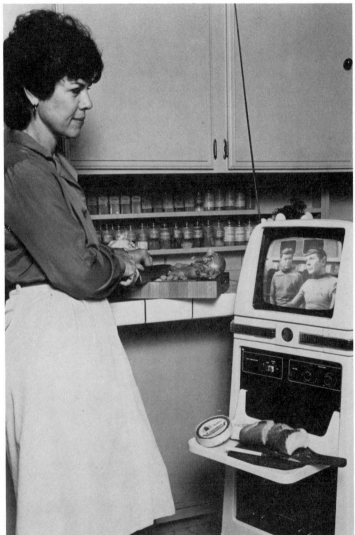

The Hubot.

psychiatrist while the user acts as a patient. The computer remembers your answers and uses them to construct more questions, sometimes so effectively that people forget they are talking to a computer, not a real person.)

The Hubot also has a built-in TV, radio and video game player. When you buy a Hubot, you can throw away your old portable TV set because you now own a *mobile* TV set! It's quite reasonable to expect that, in the near future, home robots will be able to control all the audio and video components in your home entertainment center.

When equipped with a fully functioning arm, your home robot can also play board games with you. The Novag chess robot, although not a general-purpose robot, plays chess the way a human opponent would—by physically picking up the pieces and moving them. It plays quite a good game, too.

NOVAG INDUSTRIES LTD.

The Novag chess robot.

As personal robots become more sophisticated, we can expect to see them playing all kinds of action-oriented games. Imagine an all-star football game played by two teams of robots programmed to follow the strategy of the greatest athletes ever to play in each position. What a game that would be!

Robots as Educators

Seven years ago, we had no microcomputers in the classroom. Today, you can hardly pick up a magazine or newspaper, or turn on the TV, without seeing something about computers in education or "computer literacy" as it's called. Well, "robot literacy" will be the next educational opportunity for parents and teachers to become involved in.

It doesn't take much to turn an entertainment robot into an educational robot. When the games your robot can play begin to have some educational value, your toy has become an educational tool. Almost any game can be adapted to include the use of a robot, resulting in dramatic improvements in student motivation. The RB5X from RB Robot Corp., for example, can play a mathematics quiz game.

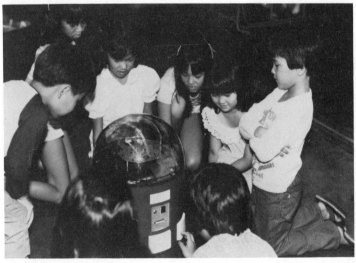

The RB5X.

RB ROBOT CORP.

One of the most popular computer languages used in schools is Logo, a language designed to acquaint children with computers and teach them the principles of computer programming. Several robot manufacturers have written software that enables Logo programs to move a robot around on the floor, just as they move a figure around on a screen. The Turtle Tot, imported into the U.S. by Harvard Associates, and Androbot's FRED are two relatively inexpensive robots intended to be used primarily with Logo.

The Logo language is based on the concept of allowing a child to get used to working with computers by using a simple set of commands to move a pointer around on a computer monitor. The pointer is called a turtle. When you type "FD 50" in Logo, the turtle on the screen moves forward 50 steps. Similarly, "BK 50" moves the screen turtle backward 50 steps, "RT 90" pivots the turtle 90 degrees to the right, and "LT 90" pivots it 90 degrees to the left.

A Logo program to draw a square would look like this:

```
FD 50
RT 90
FD 50
RT 90
FD 50
RT 90
FD 50
```

This may not look like much, but a whole new world can open up for a child who types in these simple commands and begins to understand the connection between the instruction and the result, and between a square and four lines of equal length joined at right angles. Robot turtles can be even more effective at bringing home the relationship between abstract geometrical relations and the motion of real objects in the real world.

Robots can be used to teach students about robot technology, of course, as well as computer technology. The Heath Hero 1 robot has the ability to move forward and backward, turn right and left, and move its arm. The arm has six different axes of motion—more than some industrial robot arms. The Hero 1 also has a multitude of sensors you can experiment with, including ultrasonic ranging and detection of visible light, infrared light, sound and motion. For those who want to

HEATH CO.

The Hero 1.

learn about robots from the ground up, the Hero 1 is available in kit form. You can even buy a 1,200 page robotics education course designed to complement the kit.

Teachers have had a great deal of success using robots to illustrate abstract mathematical concepts, particularly geometry. Instead of trying to hold students' attention with drawings of lines and angles, teachers can involve the students directly by having them program a robot to move a given distance then turn a certain number of degrees. This way, a child quickly learns how many 90-degree angles there are in a square. Because students acquire knowledge through direct experience, rather than by memorizing facts, they are more likely to retain the new knowledge and be able to build on it.

All personal robots are equipped with sensors for gathering information about their environment. Since these sensors work by measuring some physical quantity and converting the result into a number, they can form the basis of many interesting and informative experiments in physics. For example, the effects of atmospheric pressure on the speed of sound may be observed by placing a small robot in an enclosed area and removing the air with a vacuum pump. The operation of the robot's sonar sensor would change significantly as the air pressure begins to fall.

A talking robot can speak in French, Spanish or German almost as easily as in English. (Some of the sounds common in foreign languages but not found in English would have to be programmed in before a robot could speak without a distinctly American accent.)

With a synthesized voice of sufficient quality, a robot could be used to teach a foreign language. If the robot also had voice recognition, you could give it a word in English and it would tell you the corresponding word in French. You could even program your robot to stop speaking English with your children and respond only when addressed in French!

5

Robots as Companions, Security Guards and Servants

Personal robots have already been around long enough for familiarity to set in, at least among those who work with robots every day. But it is still surprising, and highly amusing, when you're working alone on a project (or at least you *think* you're alone) and your robot pipes up with "low voltage" or starts to sing a few bars of a popular song. A robot is, in many ways, like a pet that can talk. Robots are not warm and furry (at least, not yet) but their ability to talk, especially when you're not expecting it, is a very endearing quality.

As inanimate objects, robots should be referred to as "it" rather than "he" or "she," but it turns out to be quite difficult to avoid the more personal usage. People who don't have pets will sometimes refer to someone else's pet as "it," whereas any serious dog or cat owner will invariably use "he" or "she." Perhaps the same will hold true for people who own robots, compared with those who don't.

Robots as Companions
A robot could make an excellent pet, especially for those who don't want a pet animal or couldn't have one for some

reason. Perhaps you're allergic to cat fur, or you live in an apartment or nursing home where animals are not allowed. Maybe you'd like a dog but don't think you have time to exercise it properly.

A pet robot can provide some of the same kind of companionship you would expect from a pet animal, but without imposing any needs of its own on you. Even if you stop "feeding" it (i.e., by letting its batteries run down) you would not be guilty of cruelty to your pet. A pet robot is always there when you need it, and never in your way when you don't need it. It doesn't shed fur, "eats" only a few pennies worth of electricity per day, and doesn't need any expensive shots from the vet. When you go on vacation, you don't have to worry about who's going to take care of your pet, and there are no boarding kennel fees to pay.

There's no reason why a pet robot shouldn't be made in any shape you can think of. You could have a dog-shaped robot, one that looks like a cat or a mouse, or even a giraffe if you like. You don't even have to be restricted to real animals. Why not have a pet unicorn or a dragon?

At the moment, personal robots all look more or less robot like. But they can still function as companions, playing games, telling jokes and singing and dancing their way around your house. Even a robot that does nothing more sophisticated than answer when asked for the time can make an individual feel a little less alone. At least one manufacturer has built synthetic "emotions" into a robot. When it wins a game it dances about and looks happy. When it loses, it shrugs its shoulders and looks sad.

Robots have the potential to become highly desirable companions because, by their very nature, they never get tired or complain. Their infinite patience and complete inability to take personal offense at anything you say make them especially useful as companions for people who have great difficulty making and keeping friends. No matter how personally obnoxious an individual might be, a robot will be able to get along with him or her perfectly well.

For the elderly and the disabled, a robot companion can be

particularly suitable, because it has no needs of its own that might interfere with its job of providing around-the-clock companionship. A robot companion could be made soft, warm and cuddly so that it could snuggle up to its owner at night. As well as providing emotional support, the robot could monitor its owner's temperature and respiration throughout the night.

If the robot detected an unusual breathing pattern it could ask its owner if everything was O.K. The robot could call for help over the telephone if its owner did not respond satisfactorily. An added benefit is that a robot companion could be programmed to relay medical information to paramedics in the event of an emergency.

Robots as Security Guards

The immunity to boredom and fatigue that makes personal robots so useful as around-the-clock companions also makes them ideal as security guards. Most personal robots are equipped with sensors for detecting heat and motion, and can easily be programmed to respond to fires or break-ins. A voice synthesis system provides the robot with the ability to formulate an appropriate request for assistance and use the telephone to call for help.

If the robot detects a fire, it can issue a warning to the occupants of the house, call the fire department, and perhaps even make an attempt at putting out the fire by pointing a built-in fire extinguisher at the source of the heat.

In the case of a break-in, the robot can warn an intruder that he has been detected, sound an alarm, and call the police. An on-board video camera and recorder would be extremely useful for recording an attempted crime in progress. The robot's owners could play back the tape to find out what caused the robot to sound the alarm, and even provide the police with pictures of the intruder. Your robot could also be equipped with a voice-operated "panic button" for summoning help in an emergency.

Whenever you went on vacation, or even when you were planning a late night out, you could program your robot to run around the house turning lights and TV sets on and off, giving the house the appearance of being occupied. If, despite these precautions, someone did attempt a break-in, the robot would be on hand to sound the alarm. Any time you were away from home, you could dial your own telephone number and get a status report from your robot.

Robots as Servants

Personal robots have been described as offering us the benefits of "slavery without guilt." A survey taken among well-educated women in 1983 asked whether they would like a home robot and what they would want it to do. The results indicated that 94% of the women would like to own a home robot. They said they would use it for serving drinks, and as an answering service and burglar alarm. They also expressed interest in having the robot function as a handyman, painting ceilings, lifting heavy objects, and helping with gardening chores such as digging holes and mowing the lawn.

The survey respondents showed an admirable understanding of what a robot servant could be expected to do, namely deliver messages, run errands, and perform general housework.

A robot servant can be of real use just by reminding you of something you have to do. Something we all have to do is wake up in the morning to go to work or school. Instead of waking up to a shrill alarm or, what is sometimes worse, a clock radio announcing the latest crisis or catastrophe to strike the nation, you could program your robot to wake you up gently. If a quiet "Time to get up now" didn't do the trick, the robot could become progressively more vocal.

A robot's built-in intelligence could be expected to take care of such mundane considerations as daylight saving time changes, as well as knowing the difference between a weekday, when you have to get up, and a weekend or holiday, when you can sleep in longer if you want to.

Your robot can also remind you when it's time to turn on your favorite TV program. Maybe the robot could even record the program for you if you don't have time to watch it live. And if the kids have the TV on too loud, a robot can sense this and tell them to turn it down. When dinner is ready, instead of shouting yourself hoarse or running from room to room to call everyone to the table, let the robot do it.

With a robot in your house, you'll never again have to run to answer the phone. The robot can answer the call on the first ring, say "One moment, please," and bring the phone to you. You could even program your robot to find out what's playing at the movies. The robot would simply dial the theater and pass the recorded show times on to you.

A few department stores and supermarkets are already experimenting with shopping by means of telephone or cable TV hookups. You could improve the usefulness of such a service by having your robot keep a running inventory of groceries used and reorder as necessary. This could, perhaps, be done by installing a home version of the bar-code scanner used in most supermarkets now.

One of the classical uses for a personal robot, mentioned again and again in magazine articles, is serving drinks at a party. The robot can mingle with your guests, asking if anyone would like a drink. If someone says yes, the robot will stop to allow your guest to pick up a drink from the tray the robot is carrying. The robot will sense when the tray is empty and return to the kitchen for refills.

Another classical application, and perhaps one of the first that people think of, is vacuuming the floors. Although vacuuming seems like a simple enough job, it actually requires very sophisticated room-mapping and obstacle-avoidance systems that are just not available yet. The problems of steering around furniture and picking up loose articles lying on the floor are extremely difficult. First-generation robots may have optional vacuum-cleaner attachments, but they will be of limited use. It will be several years before robots are clever enough to do a first-class vacuuming job on an entire house.

Another highly desirable function that is not likely to be available for some time is a dishwashing capability. The kind of hand-eye coordination required for washing dishes, or even loading dishes into an existing dishwashing machine, is not yet available. It's quite likely that specially designed dishwashing machines will be built to get around the limitations of early personal robots.

Home robots will be more useful, initially, for carrying objects from one room to another around the house. Although your robot won't do your windows for you, it can at least carry all your heavy and messy cleaning materials, following you around the house as you move from room to room.

Robots never get tired or bored, so they're great for doing our tiring, boring jobs for us. With the supervision of its owner, a robot could mow the lawn, spread fertilizer, rake up leaves, and possibly even do some spray painting.

It would be asking too much to expect today's personal robots to completely replace human babysitters, but a robot could certainly assist in taking care of a baby. At the very least, the robot could listen for the baby's cries and get someone to look at the baby if the crying continued for more than a few minutes. A robot could be used as a "smart doll," although the robot might hold the child, rather than the other way around. Such a robot could also fulfill secondary roles as teacher and security guard.

One of the most useful applications of personal robots, and one that is already achievable to a certain extent, is in assisting the elderly and the disabled. A robot can help people with limited mobility, for example, by removing a heavy dish from an oven, carrying a bucket of water from one room to another, or picking up an object dropped on the floor.

The lack of highly advanced artificial intelligence software, which is holding up the development of vacuum cleaner and dishwashing robots, does not present a problem in this application. The robot can be operated by its owner, standing right next to it with a joystick control. Essentially, the robot provides the strength and physical mobility that its owner lacks, and the owner provides the intelligence that the robot lacks.

Personal robots can increase a disabled person's independence by substituting for a missing or paralyzed limb. Rehabilitation specialists in hospitals around the country are very excited about the potential for using robots to restore a measure of independence to disabled individuals.

Robots with voice recognition are particularly useful for helping quadriplegics who may have no means, other than through spoken commands, of controlling a robot. The potential benefits of personal robots in this area alone are enough to more than repay all the time, money and effort being spent on robot research.

must be equipped with better arms. A robot arm should be able to bend at several joints, grip objects firmly, and lift them even if they weigh 50 or 100 pounds.

An important concept in robot gripper technology is *compliance*—the ability to adjust to minor imperfections in the fit of the gripper around an object. Robots also need pressure sensors to prevent them from applying so much lifting or gripping force that they damage the object they are trying to pick up. These sensors, in turn, will require local processing power for interpreting the signals they generate.

Robot mobility is another function that can be improved by the addition of local processors. Just as people use information from each of the body's joints to maintain balance, a robot could control its movements by coordinating the feedback from motors and sensors located at all the "joints" in its body.

A robot with legs instead of wheels would be much better at negotiating steps and uneven surfaces. Robots could have any number of legs, although the use of two legs would offer the advantage of making a robot more human-like. One of the drawbacks of using legs instead of wheels is that more power is required to move a given distance (which is why it's harder to run 10 miles than to ride a bicycle the same distance).

But regardless of whether a robot has legs or wheels, an increase in available power would be desirable so that the robot could move faster, carry heavier loads, and run longer between recharges. New battery technology can be expected to help out here, as well as the development of sophisticated software for enabling the robot to find an electrical outlet and plug into it without human assistance.

Software development will be even more important than hardware improvements in the evolution of personal robots. Most of the hardware needed to build robots has been available for many years. The only reason personal robots have not appeared before now is that semiconductor memory and tiny, powerful, inexpensive microprocessors have only recently been introduced.

Now that we have all the hardware we need, the next problem is making it do what we want, which is where the software comes in. Software for guiding your robot around your house so that it doesn't bump into the walls, knock over the furniture, or run over the cat. Software for recognizing an object you've asked the robot to fetch for you. Software for detecting fire, burglary or burst water pipes. Software for teaching your kids Spanish, or for calling the paramedics when your 80-year-old mother falls down while she's alone in the house and can't get up. Software for all the tasks you want your robot to be able to do.

There are two schools of thought concerning the appearance of robots. One theory holds that robots are machines for performing certain tasks and that they need not look like people at all. Some people go so far as to insist that robot manufacturers should avoid any human resemblance in their products.

Probably the majority view, however, is that personal robots should look like people. There are several sound reasons for this. First of all, we have shaped our environment to match human physiology. It wouldn't make any sense to construct robots wider than our doorways or taller than our ceilings.

In any case, the human shape is quite an efficient one, with the important senses clustered in the head—at the highest point in the body, where greatest effective range can be achieved. The placement of the arms and hands also makes sense, close to and in full view of the eyes. The legs have to be at the base of the body, and this leaves the central region for the power supply (heart, lungs and digestive system) and other miscellaneous parts. There is a strong analogy between the human configuration and that of a typical personal robot.

In addition to physical considerations, there are psychological reasons for making our robots look like people. Throughout our history, we have been fascinated by the concept of creating artificial life. Look at the success of the "Frankenstein" book and the hundreds of derivative movies whose plots revolve around a scientist stumbling on the secret of how to breathe life into inanimate objects. One of the reasons personal robots are so irresistible to most people is that, primitive as they are at this early stage, personal robots already convey a strong impression of being alive.

With improvements in the physical construction of personal robots, combined with increased sensory and speech capabilities, we can expect to see future personal robots looking more and more like people.

Applications

Robots can already be used to entertain young children. Their entertainment value for older children and adults, however, is for the most part limited to the intellectual challenge of programming them. But future robots will be complete home entertainment centers. As well as being able to sing and dance and tell jokes, they will function as controllers for all your electronic entertainment equipment, such as TV, radio, stereo, computer games and telephone.

Your robot will either have this equipment built in, or it will be connected to it electronically. Just tell your robot you'd like to see the news, and it will turn on the TV, search its database for a station carrying news at this particular time of day, and tune it in. Do you have a sudden impulse to listen to Tchaikovsky's *1812 Overture* at high volume? Just tell the robot and it will select the right compact disk, switch on the hi-fi system, turn up the volume, and let it rip!

Computer games and a wireless telephone can easily be built into the robot, so all you have to do is call the robot to have these functions brought to you wherever you are. No more running out of the bathroom, dripping wet, to pick up the phone. Let the phone come to you. And if you don't want to take the call, have the robot take a message.

After entertainment, education is the next easiest application for a personal robot. After all, robots have one of the attributes a teacher needs most—infinite patience. They are capable of repeating the same material over and over again, day after day if necessary, without ever becoming tired, bored, or frustrated.

Using principles developed for computer-aided instruction, robots can adjust the pace of their delivery to each student's ability to learn. Tests can be administered and scored automatically, with automatic repetition of any material that the student has failed to absorb.

Perhaps the greatest value of robots in education is their ability to motivate even the most jaded students. With robots in the classroom and at home, our educational system may begin to recover some of the effectiveness it seems to have lost in recent years.

Entertainment and educational applications are relatively easy to build into personal robots, because, for these applications, the interface between the robot and the real world is not very complex. Tasks requiring the robot to act as a companion, security guard or servant are more difficult because of the many complex interactions necessary. But some of the applications that we can only talk about today will almost certainly become part of our everyday lives in the near future.

Personal robots could be the solution to the growing problem of caring for our elderly population. As more and more people pass retirement age and begin to need assistance in taking care of their daily needs, our social systems for helping the aged are becoming strained. Instead of dedicating one young and healthy individual to look after one elderly or infirm person, we could use robots to perform 90% of the fetching and carrying chores these people can't manage themselves.

Similarly, as an increasing number of people find themselves faced with a choice between career advancement and caring for a child, personal robots can help out. Obviously, a robot won't be able to do all the things a parent does, but at least the robot can perform some of the more mundane functions, giving parents more time to spend pursuing a career.

With advances in artificial intelligence and expert systems software, robots will eventually be able to learn, to reason, and to make decisions. This will greatly enhance their usefulness to us. It is not unreasonable to foresee an "artificial personality" to go with the artificial intelligence, making robots even more practical as mechanical pets and companions. The Heath Hero Jr. already possesses the first traces of an artificial personality.

Personal robots will definitely become sophisticated enough to perform basic household chores in the very near future. One

The super-robot could take care of all the housework, look after the children (or the grandparents), and cater to all our entertainment needs. It could be an accomplished musician, a first-class dancer and a star actor. With a sufficiently human-like appearance, the super-robot could become a sexual surrogate (in surveys, several women have spontaneously suggested this application of personal robots as a way of ridding themselves forever of troublesome boyfriends). Some people might even want to marry a robot!

The super-robot need not be limited to manual labor. With the equivalent of an entire reference library stored in its memory, the super-robot could go into the consulting business, solving complex problems or diagnosing rare diseases. (This is not just wishful thinking—expert systems programmed with the combined knowledge and experience of the world's top medical specialists can already out-perform human doctors in cases involving difficult diagnoses.)

The most significant question facing us as we consider the potential power of personal robots is this: *Can robots ever become self-aware?* Forget, for the moment, the technical difficulties involved in building the super-robot. Just assume for the sake of argument that it can be done (a reasonable assumption in view of our technological progress since the beginning of this century). What would be the implications of our ability to create a self-aware super-robot? What if the super-robot not only became self-aware but decided it wanted to run its own life without a bunch of humans always telling it what to do?

Suppose the super-robot decided it didn't want to spend the rest of its "life" cleaning the house and taking care of the kids. In the case of a self-aware super-robot, civil rights questions would certainly not be out of order. We may find ourselves obliged to protect the super-robot from total and irreversible loss of power (which, to a robot, represents life), to free it from slave labor (liberty) and allow it to choose how it spends its time (the pursuit of happiness).

These philosophical and ethical questions are very real, regardless of any personal opinions about whether or not the super-robot will ever evolve. In fact, the best answer to the old question of "What are personal robots good for?" may be that they will make us think about the nature of life and our place in the universe.

Are we trapped in the ultimate irony of working feverishly to design our own replacements as the highest form of evolution on the planet? Should we burn all robot builders at the stake as punishment for treason against the entire human race?

Let's address the second question first. It would be futile to ban the building of robots, because they can be built in a basement by anyone with a few simple tools and a basic familiarity with mechanics and microcomputer technology.

As to the first question, Isaac Asimov (the man who invented the fictional equivalent of the personal robot 40 years ago) believes future artificial intelligence will be so different from human intelligence that the two will never compete. Asimov's science fiction story, *The Bicentennial Man*, tells of a robot with capabilities similar to the super-robot under discussion here. The robot in the story becomes self-aware and develops a single, overriding ambition: to become as much like a human as possible.

No one knows how the real-life super-robot story will end. But it seems likely that, one day in the distant future, we will have an opportunity to find out.

Part II

A Brand-Name Guide to Personal Robots

7

Personal Robots: The First Generation

The personal robots we can buy today fall far short of being the all-purpose mechanical servants we would like them to be. We should remember, though, that many of the products we now use every day were so primitive that they were almost worthless when they were first introduced.

For example, the world's first automobile—a three-wheeled, steam-driven vehicle built in 1769 by Nicolas-Joseph Cugnot— had a maximum speed of only 2¼ miles per hour.

The first commercial telephone switchboard went into service in New Haven, Connecticut, in 1878, serving 21 phones. For the next 15 years, the maximum practical distance for a telephone connection was only 1,200 miles, because of signal distortion.

When the first demonstration of television was given by John Logie Baird in 1926, the monochrome picture was only a few inches high and flickered badly.

All these inventions, when they were first introduced, suffered such severe limitations that most people doubted they would ever become much more than scientific curiosities, let alone highly successful commercial products. Now, almost

Alexander Graham Bell demonstrating his telephone in 1877.

everyone in the Western world has the use of at least one automobile, telephone and TV set. Many people, especially in affluent areas of the U.S., own two, three or more of each of these products.

It's a good idea to keep in mind the humble beginnings of these now-universal modern conveniences as you read about the first generation of personal robots. You won't find any Rolls Royces here—mostly just three-wheeled machines with limited capability and a top speed of not much more than 2¼ miles per hour. Within these pages you won't see a real-life C3PO, or even an R2D2. What you will see, however, is history in the making.

The concept of owning a robot for personal use is still very new—so new that most people don't even realize yet that such robots really exist and can be purchased, taken home and put to work. But before very long, the capabilities of personal robots will begin to match our needs for mechanical servants, entertainers and companions. When this happens, personal robots will become even more widely accepted than personal computers, and we can begin to look forward to the day when there really will be a robot in every home.

8

Topo

Of all the personal robots currently available, Androbot's Topo looks most like what you'd expect a robot to look like. Topo is three feet tall, with a human-like head, shoulders and torso. Two "arms" hinge down from the shoulders and can be used to carry small objects, but they are not true robotic arms in that they cannot be moved under computer control. Topo has a unique wheel design called "Andromotion." Instead of riding on three or four wheels, Topo has two nine-inch wheels set at an angle of 30 degrees to the horizontal.

Whereas most other robots resemble R2D2 from *Star Wars*, Topo is more like C3PO. Actually, with its white plastic molded body, Topo is probably more reminiscent of the movie's Storm Troopers than either of the famous robots. In fact, Androbot's manufacturing facility in California's Silicon Valley looks more like a movie set than a factory, with dozens of identical robot bodies lined up awaiting final assembly.

Setting Up and Using Topo
Before using Topo for the first time, you have to charge its batteries. All you do is plug a small recharging unit into any

Androbot's Topo with base communicator.

standard 110-volt electrical outlet and insert a connector into a socket in the back of the robot. It's very similar to plugging in an AC adapter for a calculator or other small electronic appliance. Topo will run for about three hours on a charge, and should be left to recharge overnight and any time the robot is not in use. Topo's sealed lead-acid batteries cannot be accidentally damaged either by overcharging or excessive discharging.

To begin using Topo, all you have to do is press the power-on button. The robot will repeat its name and say "hello," then you can make it move around by pressing a four-sided switch on top of its head.

Pressing the front of the switch makes Topo move forward. The robot will continue to move until you press the back of the switch. Pressing the right side of the switch makes Topo turn to the right, and pressing the left side makes it turn to the left. You can make Topo move with curving turns to the right or left by pressing on the sides of the switch while the robot is moving forward. It's an interesting challenge to steer Topo around obstacles without stopping each time you want to change direction.

You can have a lot of fun just getting used to having a robot around the house and experimenting with Topo's head switch. But you won't be able to do anything else with Topo unless you have a home computer to use as a controller. Topo was designed to be used in conjunction with an Apple II+ or Apple IIe computer, with support for other home computers to be added later.

Using Topo with the Apple Computer

To use Topo as it was intended to be used, you'll need an Apple II+ or Apple IIe computer with a monitor and a disk drive. Every Topo is supplied with a manual, a copy of the TopoSoft software, and a base communicator.

The base communicator is a small flying-saucer-shaped object, about three inches high and six inches across. It is connected to a standard RS-232 port in the back of the Apple computer, and it works on the same principle as a remote-control device for a TV set. The computer sends instructions for Topo to the base communicator via the RS-232 port. (If you're not sure what is meant by "RS-232 port," check the glossary in the back of this book.) The base communicator converts the instructions into pulses of infrared light which are then picked up, acknowledged, and obeyed by Topo.

The infrared pulses will bounce off the walls and ceiling in your house and reach the robot even if there is a large obstruction directly in the path between the base communicator and the robot. If Topo is out of range of the base communicator (in another room or more than 25 feet away), the base communicator will report to the computer that it has lost contact with the robot.

The disk containing the TopoSoft software for controlling the robot also includes a demonstration program. To run this program, all you have to do is type the word "DEMO." Topo responds immediately with "Hello. I am your new Topo. Press my four head switches to have some fun." As you press each switch in turn, you get a brief display of singing and dancing, including four bars of "Daisy, Daisy" which seems to have become the robotic national anthem since its original performance by the crazed Hal in the film *2001: A Space Odyssey*.

Some robots use "canned" speech that works by playing back words stored in memory. Only those words that have been stored in a particular "vocabulary" can be retrieved. Topo, however, uses a highly sophisticated text-to-speech system that works by breaking down words you type on the keyboard into phonemes, or parts of words. The phonemes are then converted into individual sounds which, when strung together, produce slightly mechanical but perfectly recognizable speech.

If you type "SAY HELLO MIKE," Topo will respond with a friendly, natural-sounding greeting. The quality of the speech is quite impressive. Because there's no vocabulary restriction, you'll find you can type in any name you can think of in place of "Mike," and still get a response. This may help to make up for all the times the Murgatroyds of this world have searched fruitlessly through a thousand tourist-shop coffee mugs, looking for one with their name on it.

Topo's manual follows what has almost become a standard format for software documentation: 5½ × 8½-inch pages in a rugged three-ring binder. The manual opens up a whole new range of commands for Topo to obey. For example, the robot can be maneuvered about the house using a joystick if your computer is equipped with one. If you do not have a joystick, you can move Topo around with keyboard commands, such as:

Command	Meaning
200 FWD	Move 200 centimeters forward
100 BACK	Move 100 centimeters backward
180 RIGHT	Turn 180 degrees to the right
135 LEFT	Turn 135 degrees to the left

The pitch of Topo's voice can be varied over a range of 63 steps, as follows:

5 SET-PITCH	Set a very low pitch
60 SET-PITCH	Set a very high pitch
24 SET-PITCH	Reset to the standard pitch

You can direct Topo to follow a series of commands, one after the other, simply by typing the commands all on the same line:

90 LEFT 200 FWD SAY "HELLO MARLENE"

causes the robot to turn 90 degrees to the left, move 200 centimeters forward, and say "Hello Marlene." The only problem with this is that Topo will begin to speak immediately after beginning the 200-centimeter forward move. You will often want the robot to wait until it has finished the forward movement before making its little speech, in which case you can use the TILL-STOPPED command:

90 LEFT 200 FWD TILL-STOPPED SAY "HELLO MARLENE"

If you want the robot to return to its starting position after delivering its greeting, you might add some more commands:

```
90 LEFT 200 FWD TILL-STOPPED
SAY "HELLO MARLENE" 200 BACK 90 RIGHT
```

It would be very tiresome to have to type all this in every time you wanted the robot to repeat a certain sequence of commands, so the software allows you to save command sequences and give them a name:

```
: GREET 90 LEFT 200 FWD TILL-STOPPED
SAY "HELLO MARLENE" 200 BACK 90 RIGHT ;
```

The name of this sequence is "GREET." To assign this name to the command sequence you must type a colon, a space, the name you want to use, the command sequence, another space and a semicolon. If you put any of these items (including the colon, semicolon or quotation marks) in the wrong place, or if you omit the spaces, the software will not be able to interpret your instructions.

This can be quite a nuisance, since there are at least half a dozen ways to mistype the above command, even when you think you've finally learned how to do it right. Perhaps a later version of the software will be a little more forgiving.

One extremely useful feature of the command-saving capability is that saved commands can themselves call up previously saved commands. To understand what this means, consider the following sequence:

```
: D1 SAY "I THINK THERE IS SOMEBODY AT THE DOOR" ;
: D2 250 FWD 90 LEFT 50 FWD TILL-STOPPED ;
: D3 SAY "GOOD EVENING. STEP THIS WAY PLEASE" ;
: D4 50 BACK 90 LEFT 250 FWD TILL-STOPPED ;
: D5 SAY "YOUR GUESTS ARE HERE" ;
: DOOR D1 D2 D3 D4 D5 ;
```

Look at the last line first. When you type the command "DOOR," the software will begin by executing the command "D1." This is not a built-in command, but one you have defined yourself—a short-hand way of telling Topo to say "I think there is somebody at the door."

The other commands will then be obeyed in sequence, with the effect that the robot will walk from the living room to the front door, greet your guests and lead them back to the living room where you can look up from a book you are reading and say, casually, "Oh, there you are. Can I have Topo fetch you something to drink?"

Advanced Topo Commands

You can command Topo to move wherever you wish and say whatever you like using only the commands described above. There are some more advanced commands available, however, for exercising more precise control over Topo's movement and speech. Here are some of the advanced commands for controlling Topo's movement:

SET-SPEED	Set Topo's speed
SET-RAMP	Set Topo's rate of acceleration
RESET-MOTION	Reset speed and acceleration to factory settings
ARC	Move in a curved path
PARK	Stop immediately
RESET-TOPO	Perform a complete reset
SAVE-FORTH	Save all the command sequences you have defined onto a disk for use in the future

A Robot In Every Home

Some advanced commands for controlling Topo's speech are:

SET-PITCH	Set the pitch of Topo's voice higher or lower than normal
SET-VOLUME	Set the volume of Topo's voice higher or lower than normal
TALK-FAST	Speak faster than normal
TALK-SLOW	Speak at normal speed
TALK-LEVEL	Speak in a monotone
TALK-WAVY	Speak with normal, variable intonation
TILL-SILENT	Process no more commands until Topo has finished talking
SAY#	Speak the answer to an arithmetic problem
SAY-LATER	Remember the following words, but don't speak them until the "SAY-IT" command is given
SAY-IT	Speak the words saved under all previous "SAY LATER" commands
SAY-LETTERS	Pronounce each letter individually, rather than complete words
SAY-WORDS	Revert to the normal setting of pronouncing complete words
SAY-ALL-PUNC	Pronounce the names of all punctuation marks found in the following text
RESET-SPEECH	Reset all speech settings to standard factory settings
PHON	Pronounce the following phonetic symbols

The last command, PHON, allows you to specify precisely a series of sounds you want Topo to speak. You can use the PHON command to alter the pitch, length and loudness of individual syllables to obtain speech intonations that are much more realistic than the standard synthesized speech. You can even make Topo speak foreign languages, but the robot will have an obvious American accent because the words will still be made from the sounds found in American speech, strung together in an English rhythm.

Sometimes, because of the oddities and ambiguities of the English language, you have to alter the spelling of certain words before Topo can pronounce them correctly. Rather than being an inconvenience, this can turn into an unexpected educational bonus. A child experimenting with different ways of structuring a word to make it sound right would surely come away with a better understanding of how language works.

Notes on Topo

Topo can navigate very precisely on both bare and carpeted surfaces, but has considerable difficulty moving from one type of surface to the other. The robot tends to spin its wheels before gaining a firm grip on the new surface, causing it to lose track of how far it has travelled. A series of maneuvers intended to take Topo back to its original starting position would be unlikely to succeed if part of the journey involved a transition between carpeted and uncarpeted surfaces.

This is a problem afflicting any robot that uses wheels for mobility, and one that will not go away until methods other than counting wheel revolutions are used for robot navigation (for example, using beacons to determine exact position by triangulation).

Perhaps a bigger drawback as far as Topo is concerned is the requirement for an external computer to control the robot. Quite apart from the inconvenience and expense of needing all that additional hardware, the use of the base communicator to link the robot to the computer through infrared pulses is severely limiting. The robot can't go more than 25 feet from the base communicator, or around a corner and into another room. This means you can't program Topo to carry hors d'oeuvres from the kitchen to your guests in the living room. And the robot can't fetch you a beer from the kitchen while you sit in front of the TV watching *Monday Night Football.*

But in spite of these limitations, it's certainly more fun to program Topo to walk and talk than it is to write computer programs that can do nothing more than display the results of your efforts on a small, flat screen.

Topo is priced at $1,595 including TopoSoft software, infrared base communicator, battery recharger, and manual.

9

Hero 1

Heath's Hero 1 is a box-shaped robot, 20 inches high, about 18 inches in diameter, and weighing just under 40 pounds. The drive mechanism consists of two five-inch non-driven wheels and a single, driven, steerable wheel. The Hero 1's head can rotate, and serves as the mount for an ultrasonic ranging transmitter and receiver, an infrared motion detection transmitter and receiver, a light sensor and a sound sensor.

A keypad for communicating with the Hero 1's on-board computer is also mounted on the robot's head. The keys are labelled 0 through 9, A through F and RESET. Additional labels identify control functions associated with particular keys; for example, PROGRAM, MANUAL, LEARN and REPEAT. An ABORT key, a SLEEP switch and a six-position LED display are located near the keypad.

Two important options for the Hero 1 are a voice synthesis system and a fully functioning arm. The voice synthesis system is phoneme-based, so its vocabulary is essentially unlimited. The arm has what roboticists call "six degrees of freedom," meaning that it has six independent kinds of motion: gripper opening and closing, wrist rotation, wrist pivoting, arm exten-

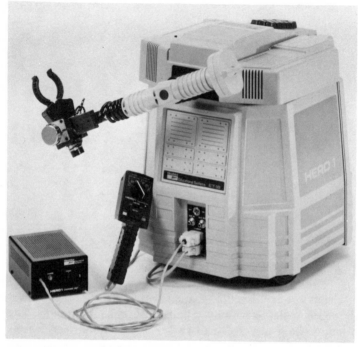

HEATH CO.

Heath's Hero 1 with teaching pendant and battery recharger.

sion, shoulder rotation and head rotation (which counts as an arm motion because the arm is attached directly to the head at the shoulder).

This robot also comes with a teaching pendant—a control box that attaches to the robot with a cable, used to "teach" the robot to perform a desired series of operations. The teaching pendant has a function switch, a rotary switch for selecting which part of the robot you wish to move, a motion switch for specifying which direction to move in, and a trigger switch to start and stop the motion. Industrial robots are often programmed using a teaching pendant very similar in function to this one.

The back panel of the robot has a socket for plugging in the teaching pendant, another for plugging in the battery recharger,

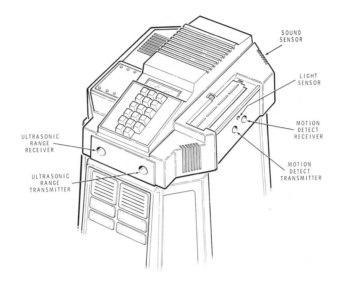

SOUND
SENSOR

LIGHT
SENSOR

MOTION
DETECT
RECEIVER

ULTRASONIC
RANGE
RECEIVER

MOTION
DETECT
TRANSMITTER

ULTRASONIC
RANGE
TRANSMITTER

The Hero 1's sensors.

an on/off switch and two jacks used for passing programs
between the robot's internal memory and an external tape
recorder.

Setting Up and Using the Hero 1

The Hero 1 is shipped from the factory with its rechargeable
battery, body panels and optional arm packed separately to
avoid damage while in transit. Installing these items is not the
most difficult job in the world, but it is a little more challenging
than just plugging components into place. You have to page
back and forth in the manual and wrestle with small Allen
wrenches and tiny nuts and bolts that are difficult to get a
good grip on. This is probably a legacy of the Hero 1's origin
as a kit-built robot.

After leaving the Hero 1 hooked up to the battery recharger
overnight, you can turn the power switch on and get the robot
to say "Ready." That's about all you can expect to accomplish

without sitting down and studying the manual. Unless you know what you're doing, you can't even get the robot to move by pushing buttons on the teach pendant.

TEACHING PENDANT

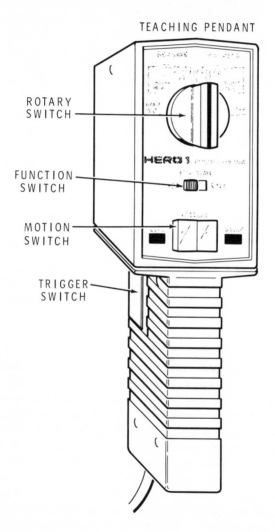

ROTARY
SWITCH

FUNCTION
SWITCH

MOTION
SWITCH

TRIGGER
SWITCH

The Hero 1's teaching pendant.

The reason the Hero 1 may seem a little more difficult to get started with than some of the other robots described in this book is that the Hero 1 was designed to be an educational robot, rather than a home companion or entertainment unit.

The Hero 1 doesn't spring to life the minute you open the box, but a few minutes spent with the manual will reward you with some insight as to the tremendous power and range of possibilities of this robot. Heath has even put together an accompanying robotics and industrial electronics course to help students and teachers get the most out of the robot.

When the Hero 1 is turned on, and whenever you press the RESET key, the robot enters what's called the "executive mode." This is the state from which you can access the other modes: the manual, program, learn, repeat and utility modes.

Manual Mode

To use the teach pendant, you first have to press the key labelled MANUAL to enter the manual mode. Then you can use the teaching pendant's function key to select either arm motion or body motion, and the rotary switch to further specify which motor you wish to control. The motion switch is used to select the direction of motion, and the trigger to start and stop the selected motion.

Learn Mode

The learn mode is very similar to the manual mode, except that the Hero 1 remembers every move you tell it to make, and can play those moves back over and over again in the repeat mode. To get into the learn mode from the executive mode, you just press the key labelled LEARN. Then you enter an address where the Hero 1 should begin storing the programming code corresponding to the movements you specify. An address such as "0100" will do the job, as well as being easy to remember.

Next, you need to enter the last address to be used in storing your program. This is to prevent a new program from overwriting and destroying an existing one. An address of "0200" will leave adequate room for a short sequence of movements.

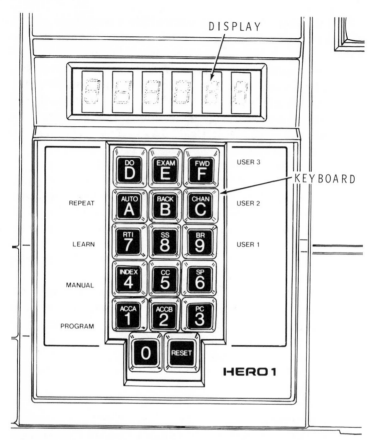

The Hero 1's control panel.

All you have to do now is use the teach pendant to move the robot around and to manipulate the arm. The Hero 1 will store the commands corresponding to each motion sequentially in memory. If you make a mistake and press the wrong button on the teach pendant, you can undo your error by pressing the BACK key on the keypad. When you press the teach pendant trigger, the robot will back out of the program it has stored by executing each command in reverse—it's like running a film or videotape backwards. When you've undone all the damage, you can press the FWD key to start moving forward again, overwriting the incorrect instructions with new ones.

There's another key labelled RTI (Reverse The Instruction) which also causes the Hero 1 to run in reverse, but this time the robot stores each reversed move. This is an easy way to get the robot out of a tight spot that you've maneuvered it into. Or you could teach the robot to load up a tray with small objects, then unload them again.

Repeat Mode

To play back the commands stored under the learn mode, you press the key labelled REPEAT, then the DO key. The display will show four dashes, prompting you to enter the starting address of the program you want it to run. (You do remember what that address was, don't you?) In the example given above, we used "0100."

As soon as you enter the final digit of the starting address, the Hero 1 will reset all motors to the position they were in when you started the learn program. The robot will then repeat all the actions in the program, performing them a little faster than when you programmed them in.

If you don't like what's happening, you can press the RESET key or the ABORT key to stop the robot.

Program Mode

This mode, entered from the executive mode by pressing the key labelled PROGRAM, allows you to enter instructions directly into the Hero 1's on-board computer. This is a very powerful mode, but there is one drawback. You have to use machine language—the internal programming language of the Hero 1's microprocessor. All instructions are entered in the form of hexadecimal codes, a form that is one step up from pure binary code, but still very tedious. (If you don't know what these codes are, don't worry—this programming mode is not something you will be getting involved with unless you decide to get really serious about computer programming. Then you'll need some really serious programming manuals.)

Heath has made things a little easier by allowing you to program the Hero 1 using the REPEAT mode rather than the PROGRAM mode. The difference between the two is that some commonly used functions have already been written for you, such as the "speak" function. If you wanted to make the robot speak using only machine-language instructions in the PROGRAM mode, you would have to use several instructions to perform this function. Using the REPEAT mode, you can make the robot speak using a single special instruction code. A built-in "robot language" feature interprets this code and automatically generates all the individual instructions you would otherwise have to program yourself.

Another convenience you get by using the REPEAT mode instead of the PROGRAM mode is automatic numbering, which saves you from having to specify the address for each instruction before you enter it into memory.

The following is a relatively simple example of a Hero 1 program to illustrate that, although the code is fairly complex, it's not impossible to comprehend, even for people who are not professional computer programmers.

Memory Address	Input	Comments
	0090	Specify program's starting address
0090	72	Speak the phonemes . . .
0091	00	. . . that start at . . .
0092	95	. . . memory address 0095.
0093	20	This command and the next one . . .
0094	FE	. . . tell the CPU to wait here . . .
		. . . when finished speaking.
0095	1B	This is the first phoneme (H).
0096	3B	Phoneme (E).
0097	18	Phoneme (L).
0098	35	Phoneme (O).
0099	37	Phoneme (U). ("HELLO").
009A	3F	"Stop" phoneme (needed at end).
009B	FF	Indicates end of phonemes.
009C	RESET	Return to executive mode.

As you can see, this is not exactly the most user-friendly programming language in the world, but it does have several advantages. First of all, you can specify precisely what you want the robot to do, step by step. Secondly, you can learn how to program in machine language (don't forget—the Hero 1 is primarily an *educational* robot).

If you make a mistake while entering a program (obviously an easy thing to do), you can use the EXAM(ine), FWD, BACK and CHAN(ge) keys to *examine* memory contents, move *forward* and *backward* in memory, and *change* instructions that are incorrect.

We won't get into the complete details on programming the Hero 1 from the keypad, because the technical depth required to do the job properly is well beyond the scope of this book. If you are really interested in this aspect of programming the Hero 1, you may wish to pick up a copy of *Hero 1: Advanced Programming and Interfacing*, by Mark Robillard (see bibliography).

Since it's so cumbersome to program the robot using the keypad, you can use a short-cut. Instead of entering instructions one by one, you can use the LEARN mode to let the Hero 1 generate most of the instructions for you. All arm and body movements can be accomplished this way, with the keypad used only to program speech. The simple "Say Hello" program given above, for example, could easily be embedded in a long, complicated sequence of maneuvers generated automatically in the LEARN mode.

Utility Mode

This mode provides some commonly used functions that you will find yourself using almost every day.

Initializing: This is a very useful function, because it resets the positions of all the robot's motors to a predefined setting. The Hero 1's arm moves to a "home" position, exercising and resetting the arm extend, shoulder rotate, wrist rotate, wrist pivot, gripper open/close and head rotate motors. The front wheel turns fully left, then returns to the center position, and the robot returns to the executive mode.

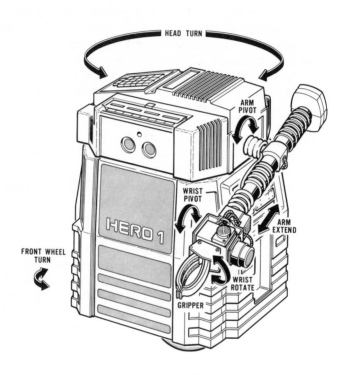

The Hero 1's arm motions.

Saving Programs on Tape: It would be very frustrating to have to reprogram your robot from scratch every time you turned the power off. Also, you would quickly use up all available memory unless you had a way to store programs for retrieval later, freeing up the memory they used. You can save Hero 1 programs on tape using any cassette tape recorder (even a $25 machine bought from a drugstore).

The robot comes with cables for connecting your tape recorder's microphone and earphone jacks to the input and output sockets on the Hero 1's back panel. To record a program, you need do little more than supply the starting address and turn on the tape recorder.

Retrieving a program saved on tape is easier still, because you don't even have to input an address.

Time and Date Functions: Other utility functions let you set and display the time and date.

Advanced Hero 1 Programming

Programming the Hero 1 from the keypad is difficult, yet potentially very rewarding. You can exercise complete control over everything the robot says and does, as well as learning a great deal about using a computer to control a machine.

Speech Control

Heath ships the Hero 1 with a dictionary of the 750 most commonly used English words. Each entry gives the codes of the phonemes that make up the word. This makes speech programming relatively easy—all you have to do is enter the phoneme codes from the book into the robot's memory. If you prefer more of a challenge, however, you can put aside the dictionary and refer instead to the phoneme charts Heath provides.

You can, for example, adjust the length of a vowel sound by experimenting with phonemes that produce the same sound, but with different durations. Placing two or three similar phonemes in sequence will tend to smooth out the sound of a word. After some practice, you should be able to create diphthongs (like the "ah" and "e" sounds in the word "I"), different accents, and changes in pitch.

Pitch changes are very important in making synthesized speech sound more natural. Consider, for example, how placing the stress on a different word completely changes the meaning of a simple sentence:

I saw him there
I *saw* him there
I saw *him* there
I saw him *there*

The Hero 1's speech synthesis system has four pitch levels for each phoneme, allowing you to build pitch variations into each word or each syllable within a word. The volume and speed of the Hero 1's voice can also be adjusted. With a higher rate of delivery, you also get an increase in overall pitch, so you can simulate the slow, low-pitched voice of a giant or the fast, high-pitched squeak of an elf.

Using the Hero 1's Senses

The Hero 1 can sense light, sound, motion, and distance to an object. The light sensor measures ambient light with a resolution of one part in 256, and stores the reading in a special cell reserved for this purpose in the robot's memory. You can program the robot to look at the value in this cell and use it to decide, for example, to turn on a light because it's getting dark.

The sound and motion sensors work in a similar fashion. You could program the Hero 1 to say "Turn the TV set down!" when the amount of noise in the room reaches a certain level. The motion detector also could be the basis for a guard function. You could program the robot to issue a challenge when it detects motion.

The Hero 1's distance measuring equipment consists of an ultrasonic transmitter and receiver. The system works by sending out a high-pitched sound signal (too high to be audible) and measuring the time it takes for an echo to be returned. A quick multiplication of the speed of sound and the time taken to receive an echo yields the distance to the object responsible for the echo. As with the light and audible sound sensors, the reading from the sonar system is stored in a special cell in memory, accessible to any programmer who knows the correct address.

Another sense the robot has is the sense of time. The Hero 1's internal clock and calendar can be accessed from a machine language program, so that you can command the robot to perform certain actions only at a particular time or on a special date. The Hero 1 can also be programmed to "go to sleep" for a specified length of time, cutting off power to almost everything except memory. You could use this feature to have the robot "wake up" every 30 seconds to check for intruders, using the motion detector. Almost no power would be used during the 29 seconds out of every 30 that the robot was sleeping.

Finally, if all this isn't enough to keep you busy, you can build your own sensors, circuitry and accessories using the Hero 1's experimental board—a small area on the robot's head designed for experimentation by advanced hobbyists. The experimental board, or "breadboard" as the hobbyists call it, carries the 12-volt, 5-volt and ground wires used in computer circuitry, and also has control lines running to and from the robot's CPU.

Notes on the Hero 1

This robot was designed primarily to be used as an educational tool, to teach students the elements of robotics. The Hero 1 packs an almost unbelievable number of features into a very small package, and it has been very well accepted by the educational community.

In view of the Hero 1's success as an educational robot, it's hardly surprising that it does *not* excel as a general-purpose personal robot. Some of the Hero 1's strengths as an educational tool translate into weaknesses when you look at the robot's performance in the more general role of a home robot.

Because the Hero 1 is not preprogrammed, and is difficult to program even if you're familiar with programming a personal computer, the robot's appeal will be mostly to individuals with considerable technical ability. It was in recognition of this fact that Heath designed the Hero Jr.—a preprogrammed robot requiring absolutely no technical ability on the part of the user.

Another limitation of the Hero 1 is the slow speed of the arm. Again, for educational purposes, the important thing is the principle of shoulder, wrist and gripper motions. If you want the Hero 1 to reach out, pick up an object, and hand it to you, it had better be a light object and you had better not be in any great hurry for it.

The Heath Hero 1 is available either in kit form or fully assembled. The difference in price is dramatic: from as little as $800 for the basic kit to as much as $2,200 for the assembled robot complete with optional arm and voice synthesis system. But this isn't a kit you can put together on Christmas morning for the kids to play with in the afternoon. Estimates for the time needed to assemble the kit range from 40 hours for an experienced electronics kit builder to over 100 hours for a complete beginner. If you wouldn't care to tackle a Heath color TV kit, think twice before you commit yourself to building the Heath robot from a kit.

Even if you take the easy way out, however, and buy a factory-built unit, you can learn a lot about robotics from experimenting with this robot. And you'll have a lot of fun while you're learning, too.

As a highly sophisticated educational package, and as an introduction to the world of robotics and a source of highly absorbing, creative enjoyment, the Heath Hero 1 is hard to beat.

10

RB5X

RB Robot's RB5X is a canister-shaped robot, 23 inches high and 13 inches in diameter, and weighing 24 pounds. The RB5X has a transparent dome on top, and moves on two four-inch wheels with two two-inch castors for balance.

Around the perimeter of the RB5X's body are eight bumper panels used both to sense contact with an obstacle and as a means of communicating with the robot (by pressing one or more bumpers with your hand). The RB5X also has a photo-diode for sensing light and a sonar system for measuring distance.

Inside the transparent dome is a circuit board on which are mounted a number of LED indicators. Four green LEDs show which way each of the drive motors is turning; eight red LEDs are linked to the eight bumper panels; and nine more light up under software control.

The back panel contains the power on/off switch, a circuit breaker reset button, a software reset button, two RS-232 ports, a socket for plugging in software modules, and a switch to tell the robot whether the module you're plugging in is a large (4K) or small (2K) module.

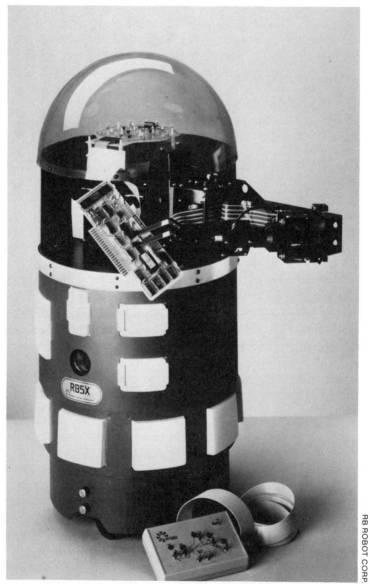

RB Robot's RB5X with arm and controller box.

Two major options you can buy for the RB5X are a voice synthesis system and an arm. The voice synthesis system is phoneme-based, and can be understood quite easily once you get used to it. The arm has limited up-and-down shoulder movement, shoulder, elbow and wrist rotation, and a gripper that opens and closes. The entire arm folds up inside the robot's body when not in use.

The arm assembly contains a lot of metal and a couple of extra batteries, both of which add considerable weight. You can install the arm yourself, but you might be better off asking your dealer to do it for you because of the arm assembly's weight and awkwardness of handling, and the four pages of detailed instructions it takes to tell you how to do the job.

Setting Up and Using the RB5X

The RB5X comes equipped with a utility software module that enables the robot to perform certain tasks as soon as it is turned on. The first thing the robot does is move forward and back, right and left, and say "Hello, I am the RB5X Intelligent Robot," as a way of checking out the motors and the voice synthesis.

Before you can do get the robot to do anything else, you need to read the manual, so this would be a good time to put the batteries on charge. The RB5X's charger is a semicircular device with two metal strips running horizontally around the inside surface. Two metal studs on the robot's body will make contact with the strips if the robot presses against the charger at the right height and at any point along the width of the device.

One of the first things the manual tells you is how to set up the charger so that the RB5X can find it and plug into it unaided. First you have to attach the charger to a wall near an electrical outlet, using adhesive tape. Then you lay a one inch wide strip of light-colored tape (if you have dark flooring— dark-colored tape if you have light flooring) leading to the charger.

When you simultaneously press bumper panels one and four (numbered clockwise, with the panel under the sonar sensor being number one), the RB5X says "Excuse me, I'm hungry," and runs off to find the charger. The robot executes a random pattern of motion until sensing the tape with its photodiode light sensor, then follows the tape to the charger. If the robot starts off in the wrong direction, it will reverse direction when it comes to the end of the tape.

If, instead of pressing bumpers one and four, you press bumpers two and five, the RB5X runs around trying not to bump into furniture or people. Whenever the robot's sensor detects an obstacle, it says "Excuse me," or makes some similar remark, then heads off in another direction.

Pressing bumpers two and four puts the robot into command mode, ready to accept a program via the RS-232 port on the back panel. The RB5X's native language is Tiny Basic, a version of the popular programming language designed to run on a simple microprocessor and take up a very small amount of memory. We will take a closer look at Tiny Basic later in this chapter.

The easiest way to program the RB5X is to plug preprogrammed software modules into the socket on the robot's back panel. Several modules are already available, and more will be added as they are developed. Some of the existing modules are:

• Voice/sound demonstration—a sampling of the RB5X's voice and sound synthesis capabilities, including speech, music, beeps and whistles, clock chimes, and an alarm.

• Intruder alarm—the RB5X sends out sonar signals and builds a "picture" of the world in front of it. You select the level of sensitivity you want and activate the alarm function by pressing various bumpers. When the RB5X's sensors detect a change in the scene in front of it, it reacts by sounding a siren, or, at your option, calling out "Intruder, Intruder!" or "Halt! Who goes there?" or "Hello, stranger."

• "Daisy, Daisy"—the ever-popular song as performed by Hal in the movie *2001: A Space Odyssey.*

● Pattern Programmer—allows you to program the RB5X to execute a series of movements. You specify forward, backward, left and right turns by pressing different bumpers.

● Bumper Music—allows you to compose a tune using the RB5X's bumpers to select notes from two octaves beginning at middle C.

● Spin-the-Robot—a robotic version of spin-the-bottle. You place the RB5X in the center of a circle of children and press a bumper to start the program. The robot spins, stops, "points" to a child and tells the child to do something, such as counting to 10 or naming all the months of the year. The RB5X then makes a comment, for example, "You're really good at this!," before spinning again and stopping at another child.

● Nursery Rhymes—All you have to do is press a bumper, and the RB5X recites a nursery rhyme.

● Carnival Barker—Intended mainly for robot dealers, this routine has the RB5X running around giving a "Come one, come all, see the fantastic robots in action!" pitch.

RB ROBOT CORP.

Plugging software modules into the RB5X.

Operating the Arm

There are three ways to operate the RB5X's optional arm accessory. The simplest is manual mode, using the controller box supplied with the arm. The box has a two-position switch for each of the five motors on the arm and two more switches for controlling the robot's wheel-driving motors.

The second way to operate the arm is in teaching mode. You start by pressing the front and rear bumpers together to initiate teaching mode. You can then move a single motor or a combination of motors, using the controller box as before. After each motion, you press a bumper to indicate that the move is complete. Pressing the right and left side bumpers signifies the end of the teaching session, and pressing the front and rear bumpers again causes the RB5X to play back the moves you have taught it.

The third way to operate the arm is by writing programming instructions for it. This is quite complicated, and will be discussed under "Advanced RB5X Programming," below.

The additional weight of the arm makes a considerable difference in the RB5X's center of gravity, so you have to be careful not to tip the robot over when the arm is installed. Also, the arm's stepper motors draw current from the battery even when they're not moving, so you need to switch the arm off when it's not in use.

Advanced RB5X Programming

To write programming instructions for the RB5X yourself, you'll need a personal computer with a software package that can transmit program files to the RB5X via the RS-232 port on the robot's back panel.

The RB5X's native programming language is Tiny Basic, which is not a particularly user-friendly language. You can make things easier on yourself by buying the optional Robot Control Language with Savvy (RCL), which uses English words instead of obscure symbols, and is very forgiving of errors such as spelling mistakes. To run RCL, you'll need an Apple II+ or Apple IIe computer with a disk drive and a serial communications card.

Let's take a look at the difference between programming in RCL and programming in Tiny Basic. Suppose you've invited some friends to your house and you want to show off your new robot. As each guest arrives, you press one of the RB5X's bumpers and the robot begins flashing its lights, runs up to your friend, stops, and says "Hello, I am the RB5X intelligent robot."

Using RCL, the code would be as follows:

```
1   Does PREPARE THE ROBOT
2   and PREPARE THE VOICE
3   and X LOAD THE INTRODUCTION
4   and CALL this robot task line "BEGINNING OF PARTY"
5   and BEGIN A LOOP
6   and EXIT IF ANY BUMPER TOUCHED
7   and REPEAT THIS LOOP
8   and TURN ON THE FLASHING LIGHTS
9   and MOVE FORWARD
10  and BEGIN A LOOP
11  and EXIT IF ANY BUMPER TOUCHED
12  and EXIT IF SONAR distance value is less than 95
13  and REPEAT THIS LOOP
14  and STOP ALL MOTION
15  and SPEAK the phrase called: "SAY THE INTRODUCTION"
16  and SPIN CLOCKWISE this many degrees: 180
17  and MOVE FORWARD
18  and BEGIN A LOOP
19  and EXIT IF ANY BUMPER TOUCHED
20  and REPEAT THIS LOOP
21  and SPIN CLOCKWISE this many degrees: 180
22  and TURN OFF THE FLASHING LIGHTS
23  and JUMP to the line called "BEGINNING OF PARTY"
24  and END
```

Although this isn't quite as easy as giving spoken instructions to a child, at least you can guarantee that your instructions will be obeyed without question! The point is, you don't need a degree in computer science to be able to understand

"TURN ON THE FLASHING LIGHTS" or "STOP ALL MOTION." The phrasing of the command "SPIN CLOCK-WISE this number of degrees: 180" is rather quaint, but at least it's clearly understandable without reference to a manual.

By way of comparison, here is the Tiny Basic translation of lines 8 through 13 of the above program:

```
670    X=#40
680    GOSUB 3100
690    @#7802
700    REM START A LOOP
710    Y=@#7800
720    IF Y<255 GOTO 770
730    D=0
740    LINK #1800
750    IF D<95 GOTO 770
760    GOTO 700
770    REM EXIT TO HERE
```

Well, that's enough of that. The entire translation runs to 115 lines of code that would be gibberish to anyone other than an experienced programmer. Compare this with the 24 lines of much more readable code in the RCL program.

Notes on the RB5X

The RB5X's limited on-board intelligence makes the robot difficult to use without extensive programming effort. Fortunately, the manufacturer has provided two solutions to this problem. One solution is aimed at young children and parents who want nothing to do with programming, and the other is designed for older children or adults who enjoy the intellectual exercise of writing their own programs.

For younger children, preprogrammed software modules enable the RB5X to recite nursery rhymes, sing songs and play games. For older children and adults, the Robot Control Language provides a welcome alternative to programming in Tiny Basic—a language more akin to machine code than to the

Basic many children learn at school or on their home computers.

In many respects, these two modes of operating the RB5X are similar to the options you have when you buy a personal computer. You can either plug in preprogrammed game cartridges, or you can learn a programming language and program the computer yourself. What the RB5X offers is a robotic version of that same concept.

The RB5X is priced at $2,295 for the basic unit, with an additional $395 for the voice synthesis option, $1,495 for the arm, and $395 for the Robot Control Language software package. The preprogrammed software modules range in price from $20 to $25.

11

Hubot

The Hubot, manufactured by Hubotics Inc., is a robot, home computer, TV set, sound system and video game machine all rolled into one. The Hubot's polyethylene body is 45 inches tall and 22 inches in diameter, triangular in cross-section, and packed from top to bottom with home entertainment features.

Starting at the top, the Hubot's head is a 12-inch black-and-white monitor that not only serves as a display screen for the robot's on-board computer, but also as a TV set. Mounted on top of the head is a button you can push to stop the robot if it's headed somewhere you don't want it to go.

Below the head is a rotating collar with flashing lights and a sonar sensor. The Hubot's chest is equipped with a time and temperature LED display, power on/off and software reset buttons, and TV tuner and volume controls. Just below these controls is a full-size computer keyboard that slides into the robot's body and completely out of sight when not in use. The keyboard is the primary means of controlling the robot, and also provides access to the on-board computer in fully CP/M-compatible personal computer mode.

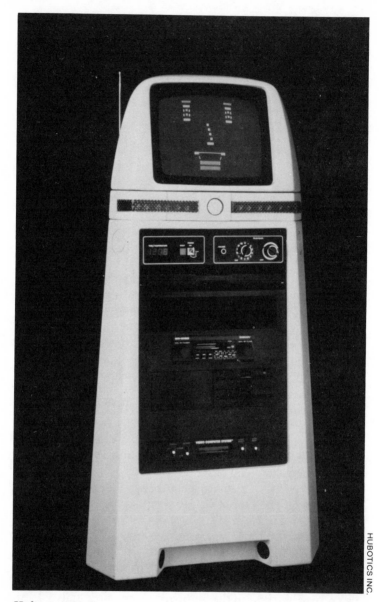

HUBOTICS INC.

Hubotics' Hubot.

Below the keyboard is a radio receiver, with built-in cassette player and equalizer, and a pair of speakers. Next to the radio package is a half-height 5¼-inch disk drive with room for an optional second drive. Near the base of the robot is an Atari 2600-compatible video game cartridge system. Finally, a serving tray can be snapped into position enabling the Hubot to carry small items such as snacks.

At the base of its body, the Hubot has three infrared sensors for detecting low obstacles. The Hubot's drive system consists of two powered wheels at the front and a single caster at the rear.

The Hubot's back panel has connectors for two joysticks (used to steer the robot and to interact with the video game system), an RS-232 interface, a parallel interface and a battery charger.

Setting Up and Using the Hubot

Getting started with the Hubot is really easy. The manufacturer provides a disk which you insert in the disk drive. When you turn the power on, the on-board computer loads the control software automatically. After a few seconds, the TV screen that forms the Hubot's head lights up with a face that moves as the robot says "Hello. I am a Hubot. How may I serve you?"

After the Hubot has introduced itself, its face is replaced on the screen by a menu of options. The main menu reads:

> Hubot can be: Smart
> Fun
> Mobile

The Smart option is the most sophisticated of the three main-menu selections, so let's start by looking at the more basic Fun and Mobile options.

Fun Option

This is the menu displayed when you select the Fun option
from the main menu:

Radio
TV
Atari
Games
Show
Talk
Lights

If you select the Radio, TV or Atari options, the Hubot's
face will appear briefly on the screen while it says "Radio (or
TV or Atari) selected." The radio, TV or Atari video game
system will then come on automatically.

If you select Games you'll get a menu consisting of the
computer games:

Hu-Libs
Tic-Tac
Eliza
Hu-Kids
Weekday

Hu-Libs is a set of word games in which you type in some
nouns, verbs and adjectives, and the Hubot makes up a story
using your words. Tic-Tac is the ever-popular tic-tac-toe,
played on the robot's display screen. Eliza is a well-known
artificial intelligence game in which the computer acts as a
psychiatrist, asking questions and using your answers as the
building blocks for further questions. Hu-Kids is a question-
and-answer game, and Weekday is a program that will tell you
what day of the week any given date falls on.

Moving back to the Fun menu, you can get a quick demonstration of the Hubot's features by selecting the Show option.
The robot's face will reappear on the screen, then the robot will
ask you to type in your name. The Hubot will say "Hello,
Mike", or whatever name you gave it, before going into its

routine. The robot will move around, turn on the TV, then the radio, then the video game system, and will then ask you to experiment with the joystick control.

To use the joystick (supplied with the robot), you plug it into a connector on the back panel, then slide the keyboard into its recessed storage compartment. As you move the joystick handle forward, backward, left and right, the robot moves correspondingly. If you steer the Hubot too close to an obstacle, it stops and says "Object detected."

The next entry following the Show option on the Fun menu is the Talk option. This selection activates the Hubot's excellent text-to-speech converter. The screen displays the words "Enter text," and, when you type in a line of text, the robot converts what you've typed into speech.

The final entry in the Fun menu is the Lights option, which lets you program the frequency and duration with which the lights around the Hubot's collar will flash.

Mobile Option

If you start again from the main menu (the one with the Smart, Fun and Mobile options), and select the Mobile option, the Hubot presents you with another menu consisting of the following options:

> Joystick
> Sonar Range
> Infrared Sensors
> Speed
> Calibration

The first option enables you to move the Hubot around using the joystick to specify forward and backward movement and right and left turns. The Sonar Range and Infrared Sensors options are for changing the sensitivity of these sensors. The Speed option, as you might guess, allows you to vary the speed at which the Hubot moves, and the Calibration option lets you make adjustments to correct small variations in the Hubot's straight-line, diagonal and rotational motion.

Smart Option

The Smart option (on the main menu) is the one that lets you program the Hubot to perform an unlimited range of tasks. The Smart menu looks like this:

Task Teacher
Task Run
Home Base
CP/M 2.2
Set Time
Voice Prompting
Sleep

Task Teacher is used to program the Hubot to do something, and Task Run to tell it to run a program you've already saved. When you select the Task Teacher, the robot asks you to provide a name for the task you're about to enter, then displays the following menu:

Move/Turn
Sonar Range
TV
Radio
Atari
Delay
Talk
Chain
Sub-Task
Lights
Collar
Schedule
Sleep
End Task

The Move/Turn option allows you to move the Hubot around using a joystick. Each motion will be stored on disk for playback later. The Sonar Range option lets you adjust the sensitivity of the sonar sensor, and the TV, Radio and Atari options are for turning each of these units on and off. The Delay option makes the Hubot pause, either for a specified length of time, or until you press a key on the keyboard. The Talk option allows you to type in the words you want the Hubot to say when it gets to this point in the task.

The Chain and Sub-Task options are used to link to other tasks already stored on the disk. This is very useful, since you only have to teach the Hubot once how to perform a commonly used set of movements and speech.

The Lights and Collar options are for flashing the Hubot's lights and turning its collar. The Sleep option deactivates the robot without actually switching off the power (so that its memory is not disturbed), and the End Task option is the one you select when you've finished defining the task and are ready to save it.

To run a task, you simply select Task Run from the Smart menu. After asking for the name of the task you want, and whether you want special options such as running in reverse or single-stepping (for debugging purposes), the Hubot will repeat the task, exactly as you taught it. With the joystick motion control and text-to-speech converter, it's really easy to get the Hubot to say and do whatever you want it to.

Other options on the Smart menu let you specify the Hubot's starting position (Home Base), run CP/M programs, set the time of day, turn off the Hubot's voice prompting, and command it to go to sleep.

Notes on the Hubot
With its built-in TV, radio and video game system, the Hubot offers much more in the way of entertainment capability than most other personal robots. But the inclusion of an easily accessible personal computer, complete with monitor, keyboard and disk drive, is what really sets the Hubot apart from its competition. You get all the power of a conventional personal computer, in addition to the entertainment systems, packaged into a robot with excellent mobility and speech synthesis capability.

The Hubot is a particularly easy robot to use, because you

don't have to learn a complicated computer language. All the robot's capabilities can be accessed by using the built-in keyboard to select options from a series of menus flashed onto the built-in monitor screen. Motion commands are intuitive, using the joystick for forward and backward moves and left and right turns. The speech synthesis system is equally straightforward—just select the Talk option and type in the words you want the robot to say.

Software for making the Hubot perform various tasks is easily transportable from one user to another, using a floppy disk. If you own a Hubot and a friend sends you a disk with some new software on it, you just enter Smart mode, pick the Task Run option, and type in the name of the new program.

The Hubot offers as much as any first-generation personal robot can offer: a good entertainment package, a complete on-board personal computer, and an extremely simple means of programming the robot to do what you want it to do. The Hubot's base price is $3,500.

Options already available or planned for introduction in the near future include a second disk drive and an on-board 40-column dot-matrix printer for use with the built-in personal computer, a voice recognition system, a sentry package with heat, smoke, and intrusion alarm, an automatic battery charging station, an articulated arm, a vacuum attachment, a remote telephone feature, and the ability to control household appliances.

12

Hero Jr.

Heath's Hero Jr. is a simplified and preprogrammed version of the company's highly successful Hero 1 educational robot. The Hero Jr. looks very similar to the Hero 1, with the same squat, boxy shape. The Hero Jr. is 19 inches high and 18 inches in diameter, and weighs 21 pounds. The drive system consists of two five-inch non-driven wheels and a single five-inch, steerable, driven wheel.

The Hero Jr. has a full complement of sensors: a visible light sensor, an audible sound sensor, an ultrasonic rangefinder, and an optional infrared motion detector. The robot is equipped with speech synthesis and has an optional radio control system.

The back panel is extremely simple, consisting only of a power on/off switch, a "sleep" switch, a socket for plugging in the battery recharger, and an RS-232 port. On top of the robot is a 6×8-inch tray for carrying small items and a cassette slot used for plugging in additional software modules. Next to the tray is a keypad and a row of LED indicators.

The keypad has keys labelled 0 through 9, A through F, ENTER and RESET. Printed above some of these keys are the names of the Hero Jr.'s built-in programs: PLAN, SETUP, DEMO, GUARD, ALARM, SPEAK, GAB, POET, SING, PLAY and HELP.

HEATH CO.

Heath's Hero Jr.

Setting Up and Using the Hero Jr.

To get started with the Hero Jr., all you have to do is take the robot out of the shipping box and slide the power switch to "on." Assuming there is still some charge left in its batteries, the robot will immediately go into its start-up routine, speaking in an easily understandable synthesized voice:

"Checking memory . . . checking sonar . . . please wave your hand in front of my sonar . . . sonar OK . . . checking steering . . . steering OK . . . checking light sensor . . . please wave your hand in front of my light sensor . . . light sensor OK . . . checking drive motor . . . drive motor OK . . . checking sound sensor . . . sound sensor OK . . . checking motion detector . . . please wave your hand in front of my motion detector . . . motion detector OK."

With all the tests completed, the Hero Jr. asks you to set the time and date. The robot is preprogrammed to change the time automatically when daylight saving time begins and ends each summer, so you must specify whether or not you are starting in the daylight saving time period.

Finally, the robot asks you to key in the date of a special occasion (a birthday, for example). With everything set, the Hero Jr. formally introduces itself:

"I am Hero Jr., your personal robot. I am your friend, companion and security guard."

The set-up process is quite enjoyable, and much easier than setting the date and time on a digital watch, but still quite time-consuming. You wouldn't want to repeat this every time you turned the robot on, and that's where the "sleep" switch comes in. When you slide the switch to the "sleep" position, you immobilize the robot without losing the date and time settings stored in its memory.

You almost don't need any instructions to run the Hero Jr. All you have to do is press the key corresponding to the desired preprogrammed function. Some of the programs have a number of options to choose from, however, so it helps to have the manual handy so that you can refer to the descriptions of all the options.

HEATH CO.

The Hero Jr.'s control panel.

The HELP Program

The HELP program is the easiest program to use. All you do is press the HELP key and the key for the program you want some information on, and the Hero Jr. will say a few words describing the selected program. If you press HELP and then HELP again, to request help on running the help program, the Hero Jr. says "You must be kidding." After a few minutes of playing with this robot, you'll discover that it has quite a good sense of humor.

The SPEAK, SING, POET, GAB and PLAY Programs

The SPEAK, SING, and POET programs are also very easy to run. When you press one of these keys, the Hero Jr. asks you to enter a number. The robot then speaks a phrase, sings a song, or recites a poem selected on the basis of the number entered. The GAB and PLAY keys work in similar fashion. Pressing GAB causes the robot to rattle off a random sequence of syllables, and PLAY directs the Hero Jr. to select a game to play, such as Cowboys and Robots—a robotic version of Hide and Seek.

The SETUP Program

One of the most unusual features of the Hero Jr. is that it has a "personality" definable by its owner. When the robot is not busy running a program at your request, it decides for itself what to do next, and just goes ahead and does it. It may sing a song, say something funny, recite a poem, play a game or go off exploring on its own. You can influence the probability of its picking a particular activity by using the SETUP command.

When you press SETUP, the Hero Jr. tells you the current settings of its "personality traits." The factory settings are as follows:

Trait	Level
Sing	1
Speak	2
Poet	1
Play	1
Explore	1
Gab	0

The effect of these settings is to make the robot twice as likely to choose the SPEAK function as the other functions (with the exception of GAB, which will never be chosen because its level is set to zero). If the Hero Jr. happens to choose the "Explore" function it will say: "I think I will explore," then start running around the room using all its sensors to try to avoid bumping into things. If the robot runs into an undetected chair leg it might say something like "Oops," "Excuse me," or "Who put that there?"

After the robot tells you what the current personality settings are, it gives you an opportunity to change each setting. The combined effect of specifying the probability that each function will be chosen defines your robot's own individual "personality."

The GUARD Program

The GUARD program commands the Hero Jr. to function as a security guard. You can tell the robot to stand guard in one spot or to move around, using its sensors to detect motion. The robot asks you for a password (which you supply by clapping your hands several times). When the robot detects motion, it says "Intruder Alert! Intruder Alert! You have five seconds to identify yourself." If you supply the correct number of handclaps, the robot says "Alert aborted. You may pass, friend."

If you fail to give the password, the robot says "Password incorrect. You are an intruder. I have summoned the police," and uses its voice synthesis system to generate the sound of an alarm going off. This is not an empty threat, by the way. If you install an appropriate Heath security system in your home and equip the robot with a special transmitter, you can program the Hero Jr. to set off your main household alarm system and alert neighbors to call the police.

The ALARM Program

The ALARM program has nothing to do with security. This program lets you put the robot to sleep until a specified time, or program it to function as an alarm clock, using the keypad to enter the time when you want the alarm to go off. If, for example, you select 7:30 in the morning, the Hero Jr. will wake you up by calling out cheerfully: "Good morning! It's seven-thirty A.M. Please wake up! Please wake up! Are you awake?"

If you don't acknowledge by pressing the RESET key, the robot says "Well, I'll let you sleep for ten more minutes. I hope you're not going to be late." The Hero Jr. will make two more attempts to wake you up, at 10-minute intervals, before giving up on you and announcing: "This will be your last call."

The ALARM program can also be used to program the robot to remind you that it's time to do something (for example, to turn on the TV to catch a movie). A clever feature of the ALARM program is that it adjusts its message, using "Good morning," "Good afternoon," or "Good evening," depending on the time of day—definitely a nice touch.

The PLAN Program

The PLAN function allows you to program the Hero Jr. with as many as 20 different tasks chosen by you, to be performed at the date and time you specify. When you press the PLAN key, the Hero Jr. will say "Begin plan" and ask you to enter the date and time you want it to perform the tasks you are about to assign. If you just press ENTER instead of keying in numbers for the day, month and year, the robot will interpret this as an instruction to perform the requested tasks at the same time every day.

After you enter the time, the Hero Jr. will ask you to confirm that you entered the date and time correctly, and will then say "Next." This is your cue to press one of the program keys. Instead of running the selected program immediately, the robot will store the request in memory. You can then select another program key, and keep repeating this cycle until you press the PLAN key again to indicate that you have no more tasks you wish to enter. The Hero Jr. will respond with "End plan."

When the date and time you specified comes around, the robot will automatically perform the tasks you requested. This is a good example of providing the user with the ability to program a robot without first having to learn a complicated programming language.

The Demo Program

The DEMO program presents a demonstration of many of the Hero Jr.'s capabilities, including speech, sensing light and sound, measuring distance and acting as a security guard.

Remote Control Option

The Hero Jr.'s optional radio control system enables you to move the robot around under remote control. The control unit has four buttons on it. The first button controls forward and backward motion, the second and third buttons steer the robot left and right, and the fourth button returns control to the robot's on-board keypad. It's quite easy to move the Hero Jr. from one room to another using the radio control system. The robot's five-inch wheels cope quite well with the transition from hard floors to thick rugs.

Advanced Hero Jr. Commands

Although the Hero Jr. is intended primarily to be used as a fully preprogrammed robot, you can, in fact, "get inside" and change the programming if you really want to.

The trick to reprogramming the Hero Jr. is to press the "0" key while holding down the ENTER key. The robot will respond by saying "Robot Wizard," and will then wait for you to enter the address of a location in the robot's memory. Depending on which address you enter, you can perform several different kinds of wizardry. You can:

- change the robot's name from Hero Jr. to something else
- command the robot to use your name as the name of its master
- command the robot to measure the distance to the nearest object in front of it, and tell you what that distance is
- make adjustments to the robot's light and sound sensors
- transfer program files back and forth between the robot and a personal computer

A programmer's guide supplied with the robot includes a list of the most common personal names together with the codes of the phonemes the Hero Jr. needs to pronounce each name correctly. The guide also includes lists of all available phonemes, organized both alphabetically (by phoneme symbol) and numerically (by phoneme code number). With these lists you can make the robot say just about anything you wish.

It is possible to develop programs on a personal computer and download them into the robot, but you must remember that the Hero Jr. was designed as a preprogrammed robot, and reprogramming it is not easy. You would have to write your programs in assembly language and transfer the program files from the computer to the robot using a terminal emulation software package (no software or detailed instructions for doing this are supplied with the robot). In addition to all this, you have to practically disassemble the robot before you can change the RS-232 port's data transmission rate.

If the above paragraph sounds like gobbledygook to you, it's a good indication that you should forget about reprogramming the Hero Jr. yourself.

Notes on the Hero Jr.

The way the Hero Jr. works, straight out of the box, is very impressive. The biggest problem with personal robots has always been answering the question "But what can they do?" The Hero Jr. shows you what it can do as soon as you switch it on, without any need to write programs or even read the manual. This robot is designed to be operated by a child—any child, not just one knowledgeable about computers.

And that, of course, is also the Hero Jr.'s limitation. Whereas a young child will be as enchanted by the hundredth recital of "Mary Had a Little Lamb," as by the first, older children could easily become bored. For a teenager, you need a robot that can be programmed to move around the house and manipulate objects in the house. The Hero Jr.'s preprogramming does not include programmed motion (just random motion), and the robot lacks any kind of arm.

The Hero Jr. sells for $1,000, plus another $250 for the optional infrared motion detector and radio control system.

13

Turtle Tot

The Turtle Tot is manufactured by Flexible Systems, of Hobart, Tasmania, and imported into the U.S. by Harvard Associates. The Turtle Tot bears a striking resemblance to a real turtle, with a 12-inch diameter transparent dome for a shell, a pair of 2½-inch wheels for legs, and a pair of red LED "eyes" visible inside the dome. A wooden block helps the Tot balance on its two wheels, each of which is driven independently so that the Tot can move forward and backward, turn right and left, and rotate in place.

Also visible inside the dome are four contact switches which let the Tot know when it has bumped into something, and a circuit board that serves as the Tot's brain. A pen holder mounted in the center of the Tot's body will take just about any kind of pen, which it uses to trace its movements on a sheet of paper placed on the floor or table-top beneath it. To further enhance the turtle-like appearance, a "tail" extends from the rear of the Tot. The tail is actually an electrical cord used to provide power to the Tot and to transmit instructions to it from a personal computer.

When an optional speech synthesis chip is installed, the Tot can talk and sound a two-tone horn.

HARVARD ASSOC.

Flexible Systems' Turtle Tot.

Setting Up and Using the Turtle Tot

The Turtle Tot is supplied with a long power/communication cord, control software on a floppy disk, a manual, and a 10-color pen set. Before you can begin to use the Tot, you'll also need a 12-volt power supply and a personal computer with an RS-232 serial port and a disk drive. A serial interface card for use with Apple II computers and a suitable power supply can be purchased from a computer dealer or direct from the Tot's manufacturer or sales representatives. Easy-to-use control software is available for IBM PC and Apple II, IIe and II+ computers, although the Tot can be driven by a Basic program running on any personal computer that has a serial interface.

To use the Tot with an Apple II+ computer, all you have to do is connect the Tot's communication cord to the serial interface, plug in the power supply, and insert the control software disk into the disk drive. When you turn on the computer, the control software is loaded automatically and a title page comes up on the computer monitor. Next comes a menu screen that no child born in the computer age would have any difficulty using. The options presented are:

1. Controlling your Turtle with Turtle Tot Talk
2. Exercising your Turtle Tot's features
3. Programs to run with your Turtle Tot
4. Controlling your Turtle Tot with Logo

We'll start with the two easiest options, then move on to the more difficult ones.

Exercising the Tot's Features

This option requires no knowledge or programming ability whatsoever—it's purely a demonstration of the Tot's basic capabilities. Selecting this option brings up a new menu with seven more options:

A. Move
B. Eyes
C. Touch Sensor
D. Pen
E. Horn (optional)
F. Speech (optional)
G. Return to Beginning

The Move option introduces you to the Tot's ability to move around by causing the Tot to roll a short distance forward and backward, and to turn right and left. When you select the Eyes option, the Tot's eyes blink on and off. The Touch Sensor option causes the Tot to move away from the side on which you tap its shell, and the Pen option makes it draw an arrow if you insert a pen in the pen holder and place a sheet of paper under the Tot. The Horn and Speech selections instruct the Tot to sound its horn and count down from ten to zero if the speech capability is installed.

Programs to Run With Your Turtle Tot

The options on this menu are:

        A. Instant Turtle Tot Control
        B. Real-Time Instant Turtle Tot Control
        C. Morse Code Turtle
        D. Pushing Turtle
        E. Learning Turtle
        F. Talking Keyboard
        G. Spelling Turtle
        H. Return to Beginning

Instant Turtle Tot Control caters to the need for instant gratification that we all seem to have developed somehow over the last few years. By pressing only a single key, the Tot can be instructed to move *F*orward six inches, *B*ackward six inches, pivot *L*eft 45 degrees, pivot *R*ight 45 degrees, move the pen *U*p, move the pen *D*own, make some *N*oise (by sounding the horn), *W*ink, or *Q*uit (i.e., return to the previous menu).

Real-time Instant Turtle Tot Control is very similar, except that each command is obeyed continuously, rather than just once, until another command is issued. The Morse Code Turtle program lets the Tot communicate by flashing its eyes and sounding its horn in morse code.

The Pushing Turtle program is fascinating to watch. The Tot pushes an object placed in front of it, turning to adjust its position whenever the object slips off to one side. It's like watching a puppy push its dinner bowl across the floor. In Learning Turtle, the Tot quickly learns that a tap on the back means it is about to bump into something and had better turn away.

The Talking Keyboard program has the Tot speaking the names of the letters and numbers on your computer keyboard as you press each key. In Spelling Turtle, the Tot gives you a spelling test. The Tot asks you to spell several words and tells you whether you answered correctly.

Now let's take a look at some more advanced Turtle Tot programming.

Controlling your Turtle Tot with Logo

Option number 4 from the main menu is "Controlling your Turtle Tot with Logo." In order to use this option, you have to buy a copy of the Logo software for your particular computer. A description of Logo and an explanation of its usefulness in educational applications was presented back in Chapter 4, "Robots as Entertainers and Educators." But to recap briefly, Logo allows children to get used to working with computers by using a simple set of commands to move a pointer around on a computer's display screen. When you control your Tot through Logo, the robotic turtle sitting on the floor takes the place of the pointer on the screen. The Logo commands for moving forward and back, and turning right and left, will cause the Tot to perform those motions.

With the additional Logo commands PENUP and PEN-DOWN, the Tot can be made to trace its movements on a large sheet of paper taped to the floor. Other Logo commands designed specifically to be used with the Turtle Tot are EYES-ON and EYESOFF which turn the Tot's eyes on and off, and SAY, which makes the Tot talk.

Turtle Tot Talk

Even if you don't have access to the Logo software, you can still perform many of the same functions by using Turtle Tot Talk, selection number 1 on the control software main menu. The Turtle Tot Talk commands are of the form:

Command	Meaning
&F 50	forward 50 steps
&B 50	backward 50 steps
&R 90	pivot right 90 degrees
&L 90	pivot left 90 degrees
&U	pen up
&D	pen down
&E	turn eyes on
&O	turn eyes off
&S 75	say word no. 75 from the word list (which is the word "control")

There are commands in both Turtle Tot Talk and Logo to find out whether one of the Tot's four bumper switches is being pushed. With these commands, you can program the Tot to make decisions based on whether or not it has bumped into something.

Programming the Tot in Its Native Language

Turtle Tot Talk is not actually the Tot's native language. In fact, the Tot's native language is nothing but numbers. You can program the Tot from any language by accessing your computer's serial port and supplying the code number for each Tot operation.

For example, to make the Tot to say the word "control" using Basic on an IBM PC, the command would be

OUT 1016,75

"OUT" is the IBM PC's output instruction, "1016" is the designation of the computer's serial port, and "75" is the code for the word "control." Similar codes exist for the other 143 words and sounds in the Tot's speech synthesis vocabulary, forward and backward moves, left and right turns, eye and pen commands, and bumper switch interrogation.

Now that you know how the Tot's native language works, you can see how the other ways of programming the Tot were designed. Essentially, all Turtle Tot commands eventually boil down to a series of commands instructing your personal computer to send code numbers via the computer's serial port to the Tot. At the highest level, you can just make a selection from a menu and the software will do the rest for you, and at the lowest level, you can write a program to issue output commands one by one.

Notes on the Turtle Tot

The Tot is a particularly intriguing type of personal robot. On the one hand, it's relatively inexpensive at $300 for the basic unit. On the other hand, by the time you've added a serial interface and power supply ($100) and the voice synthesis option (another $100), you've got $500 invested in a 12-inch device with not much more to it than a circuit board, two wheels, four contact switches and a pen actuator.

But the value of the Tot should be measured in terms of its capabilities rather than its components. The Tot won't wake you up in the morning and make you a cup of coffee, but this little robotic turtle can teach you or your young child a great deal about programming computers, understanding fundamental geometric concepts, and controlling robots.

Since this robot is such a simple one, it is easy to learn what the robot does and how it works. Through the transparent dome, you can see the motors driving the wheels and you can watch as the contact switches are pushed in by the turtle's shell. There is only one circuit board inside the Tot, so it's obvious that this board is controlling the Tot by means of the nerve-like wires running from the board to each motor and sensor.

Finally, one of the best features of the Tot is its complete range of software, from fully automatic demonstration routines to machine-code-like programming for people who really enjoy programming robots at the most fundamental level.

The Tot does have a tendency to snag the edge of a sheet of paper when moving onto it, a problem that could probably be alleviated by adding a caster in place of the wooden block used to help the two-wheeled turtle keep its balance. Another convenience would be an infrared link in place of the long communication cord that can get tangled up when the Tot makes a series of twists and turns. One more nice feature would be some local storage so that control programs could be downloaded into the Tot and run without any connection to an external computer.

Even without these improvements, the Turtle Tot is a very workable personal robot, especially in educational applications.

14

FRED

Androbot's FRED looks like a miniature Androbot Topo robot, 12 inches high and weighing five pounds. FRED has a Topo-like head mounted on a rectangular base, with a pair of non-functional eyes and a mouth that contains a speaker. An infrared receiver is mounted on top of the head, and a pair of LED indicators set into the head light up whenever a command is received.

FRED's drive system is a single, centrally located, steerable drive wheel with two floating ball bearings at the rear for balance. A pen mechanism at the front of the robot is flanked by two small supporting wheels. The pen mechanism can be lowered below the normal pen-down position so that the collar of the pen holder becomes the pivoting point during turns. This is a neat solution to the problem of keeping the pen aligned during turns.

FRED comes with a hand-held controller with an infrared transmitter for sending commands to the robot. The controller has 25 buttons and an LED indicator that lights up when commands are being transmitted to FRED.

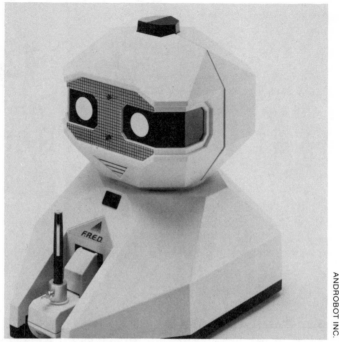

ANDROBOT INC.

Androbot's FRED.

Setting Up and Using FRED

FRED is powered by four D-size and six AA-size alkaline batteries. Before you can begin using FRED, you'll have to buy these ten batteries plus a 9-volt battery for the controller. To get the most out of FRED you'll also need a felt-tip pen. If you don't want to keep buying replacements for the alkaline batteries, you can buy an optional battery charger and use nickel-cadmium rechargeable batteries instead.

To install a pen in FRED's pen mechanism, you simply drop the pen into the pen collar and clamp it in place by tightening a wing nut.

You can check that the infrared communications are working by pressing the reset button on the controller. FRED will respond by saying "I'm FRED." Next, you can experiment with the motion controls. For example, you might enter the following sequence of commands:

FWD 2 0 ENTER

FRED will echo each part of the command, saying "Forward two zero go" ("go" being the echo for the ENTER button). As soon as the ENTER button is pressed, FRED will start to move forward (travelling a distance of 20 centimeters in this example).

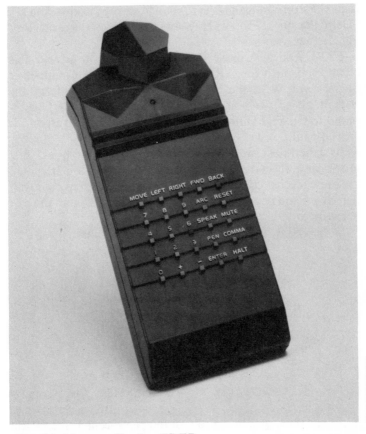

Hand-held controller for FRED.

The BACK button makes FRED go backward instead of forward. To turn left or right 90 degrees, you press

LEFT 9 0 ENTER
or RIGHT 9 0 ENTER

If you see FRED heading for an obstacle, you can press the HALT button to make the robot stop. When you press the HALT button a second time, FRED will resume the interrupted command. If FRED runs into an obstacle before you issue the stop command, the robot will stop on its own and say "Help, I'm stuck!" You will then have to remove the obstacle or enter a new command.

FRED can also move in a curved path. To achieve this effect, you push the ARC button and enter the number of degrees you want the drive wheel to turn away from the straight-ahead position. Then you use a regular FWD command to tell FRED how many centimeters to move along the curved path.

To experiment with the pen commands, you should first tape a large sheet of paper to a table-top or the floor, then use the same motion commands described above interspersed with pen-down and pen-up commands (press the PEN and the + or – buttons).

For example, to draw a square you would enter:

PEN	+	ENTER
FWD	10	ENTER
RIGHT	90	ENTER
FWD	10	ENTER
RIGHT	90	ENTER
FWD	10	ENTER
RIGHT	90	ENTER
FWD	10	ENTER
PEN	–	ENTER

If something gets in the way of the pen mechanism, FRED will say "Pen is stuck."

FRED's built-in echoing of commands and the robot's little messages such as "Help, I'm stuck!" are quite impressive, but the ability to synthesize a virtually unlimited range of phrases and sentences is even more impressive. FRED has a vocabulary of 52 words, each accessible by pressing the SPEAK button and entering the code number for that particular word.

There's a COMMA button on the controller for entering the commas separating the code for each individual word. FRED's complete vocabulary consists of the following words:

1	FRED	27	Draw
2	Go	28	Speak
3	Left	29	Learn
4	Right	30	Is
5	Forward	31	Stuck
6	Back	32	Robots
7	Turn	33	Are
8	Help	34	Fun
9	Can	35	Cute
10	Teach	36	I'm
11	One	37	Tired
12	Two, To, Too	38	Bored
13	Three	39	Funny
14	Four, For	40	Smart
15	Five	41	Good
16	Six	42	Patterns
17	Seven	43	Lines
18	Eight	44	Likes
19	Nine	45	Humans
20	Zero	46	Not
21	Units	47	Say
22	Degrees	48	Stop
23	Pen	49	Done
24	Up	50	Children
25	Down	51	I.R.
26	Move	52	Arc

Using just these 52 words you can make up a lot of useful sentences, such as:

SPEAK 1, 9, 8, 50, 29, ENTER

(which translates as "FRED can help children learn").

You can combine motion commands with speech commands in any order. For example, if you enter:

PEN +, FWD 20, SPEAK 1, 44, 12, 27, ENTER

FRED will lower the pen, draw a line 20 centimeters long, then say "FRED likes to draw."

As well as knowing how to talk, FRED also knows how to keep quiet. When you press the MUTE button, FRED stops echoing commands as you enter them (until you press MUTE a second time).

Each time you enter a command on the controller, FRED not only executes the command but also stores it in an internal memory. There's room for up to 64 motion commands, 120 speech commands, or a combination of the two. Pressing the RESET button erases this memory, as well as stopping FRED, centering the drive wheel and making the robot say "I'm FRED." You can replay all the commands stored in memory by pressing the MOVE button followed by ENTER.

It's unlikely that you could enter a complete program through the controller without making some mistakes. You can correct these mistakes by using combinations of the MOVE, +, –, and numerical buttons to jump around within the command memory, deleting commands and inserting new ones.

Finally, you can calibrate FRED to correct for minor manufacturing differences and alignment changes due to wear. Four calibration adjustments are available, correcting for straightness of motion in the forward direction, straightness of motion in the backward direction (which is different because of motor torque), angular accuracy and distance accuracy. With these calibration factors, you can almost always obtain round circles and square squares.

Using FRED with Logo

If you have an Apple II or Commodore 64 personal computer with Logo software, you can drive FRED from Logo by buying Androbot's optional Logo extension kit. The kit consists of Logo extension software, a manual and a cable. Logo commands are sent to FRED's hand-held controller via the cable, then transmitted to FRED using infrared communications.

This opens up a whole new world of operations for FRED, since anything you can do with the Logo "turtle" on the screen can also be translated into motion and speech for FRED to execute. (See Chapter 4 for complete details of how Logo works and why it's so useful in educational applications.)

Notes On FRED

FRED's infrared control link and on-board batteries eliminate the need for the "umbilical cord" that most turtle-type robots need. This frees the robot from the danger of tripping over its own cord, at the expense of a constant need for battery replacement or recharging.

Because FRED is designed with a single, central drive wheel, curves and turns tend to be more precise than with a two-wheel-drive turtle. The location of the pen, in front of the robot, does have an unfortunate consequence, however. The robot tends to obscure the drawing it is working on most of the time, so that you can only see the results of its efforts after the drawing is completed. It would have been more practical to position the pen at the back of the robot instead, so that the entire drawing could be seen as it was being made.

The ability to use the hand-held controller as a transmitter for commands coming from a personal computer is a very nice touch. You can use FRED straight from the box, even if you don't have a personal computer, and you can expand the robot's capabilities by hooking up to a computer whenever you're ready.

FRED is priced at $499, with an additional $50 for the optional battery recharger, and $79 for the Logo extension software, manual and cable.

15

Dingbot, Verbot
and Omnibot

Tomy Corp. sells three toy robots called the Dingbot, the Verbot and the Omnibot. The three robots span the range from extremely simple to very sophisticated. We'll take a detailed look at each model, starting with the simplest.

The Dingbot
The Dingbot is a four-inch-tall toy robot with a head like E.T.—complete with bulging eyes—mounted on a long skinny neck. The Dingbot has two manually adjustable arms that can be twisted into place for holding a miniature printed floor plan. The drive system consists of a pivoting drive wheel and two non-driven wheels.

The Dingbot runs on a single AA-size battery (not included), and its only control is an on/off switch. When you switch it on it runs along, chattering to itself, until it bumps into an obstacle. Then it turns and tries another direction, stopping every once in a while and turning its head from side to side as if studying a floor plan it carries.

The Dingbot doesn't have much in the way of intelligence, but this toy robot is a lot of fun and only costs about $10.

Tomy's Dingbot, Verbot and Omnibot.

The Verbot

The Verbot is a nine-inch-tall toy robot that responds to voice commands. Its arms can move up and down on command, and its eyes and mouth light up when you tell it to smile. The Verbot's drive system consists of two driven wheels and a ball bearing for the third support point. On its chest are eight buttons corresponding to each of the eight functions the Verbot can perform. The functions are: stop, smile, arms up, arms down, move forward, move backward, turn left and turn right.

Setting Up and Using the Verbot

Before you can start playing with the Verbot, you have to load batteries into both the robot and the voice transmitter used to control the robot. Batteries are not included when you purchase the Verbot, and this robot needs a lot of them. For the robot's drive motors and memory you'll need two C-size and four AA-size batteries, and for the voice transmitter you'll need a miniature 9-volt battery.

Once all the batteries are installed, you can begin to train the Verbot to respond to your voice. You do this by pressing each of the eight control buttons in turn while speaking an appropriate command word into the voice transmitter. For each command, your voice is analyzed, converted into digital form and stored in the Verbot's memory.

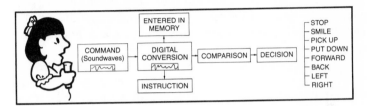

The Verbot's voice digitizing process.

To get the robot to perform one of its eight functions, you simply speak the appropriate command word into the voice transmitter. As you give each command, your voice is again analyzed and digitized, and compared with each of the voice patterns stored during the training phase. When a match is found, the robot obeys the command.

One of the most interesting features of the Verbot is that you're not restricted to any predefined command words. You can pick any word or words you like, provided the command takes between 0.2 seconds and 1.2 seconds to say. Also, you should avoid using similar-sounding words for different functions. A typical command set might be:

Function	Command
Stop	Hold it!
Smile	Smile
Raise arms	Lift up
Lower arms	Put down
Move forward	Go straight
Move backward	Backward
Turn left	Turn left
Turn right	Right turn

The Verbot interprets commands correctly about 75% of the time, either ignoring the command or executing the wrong function the rest of the time. The interesting thing is that the robot's wrong choices are amusing rather than annoying. It's like a puppy that's eager to please but sometimes performs the wrong trick. If you repeat a misinterpreted command, the chances are that the Verbot will eventually get it right. If you consistently have problems with a particular command, it's best to retrain the robot, perhaps choosing a different command word.

The Verbot's instruction book contains some suggestions for games you can play with the robot, including moving it around an obstacle course and trying to get it to obey command words stored by another player. The Verbot does not have a "sleep" switch, so you have to retrain it each time you turn off the power. This doesn't take long, however, and it's part of the fun of playing with it.

The Verbot is available in toy stores for around $60.

The Omnibot

The Omnibot is a 15-inch-tall toy robot with large, flashing red eyes, a built-in digital clock with three separate alarms, and a built-in cassette tape deck and speaker. The Omnibot's arms can only be moved manually, at the shoulder, elbow and wrist. The hands can grip small objects and hold a drinking glass or pencils and pens. Both hands can be used together to hold a serving tray for carrying objects too large to fit into either hand alone.

The Omnibot is very stable, with two pairs of rubber-treaded driven wheels, a pair of non-driven wheels, and a castering wheel. The robot comes with a radio-controlled "master control unit" that uses a joystick for forward and backward motion and left and right turns. The control unit also has a built-in microphone with a push-to-talk switch, two sound-effects buttons and a tape start/stop switch. A cardboard home-position marker is supplied to provide a fixed starting position for the robot.

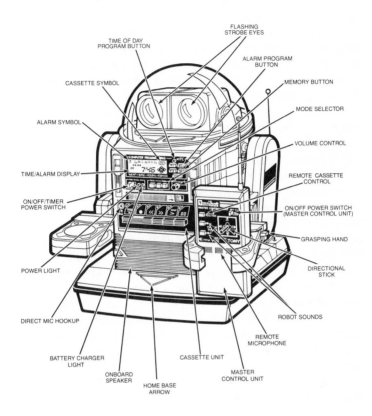

TIME OF DAY
PROGRAM BUTTON

FLASHING
STROBE EYES

ALARM PROGRAM
BUTTON

CASSETTE SYMBOL

MEMORY BUTTON

MODE SELECTOR

ALARM SYMBOL

VOLUME CONTROL

TIME/ALARM DISPLAY

REMOTE CASSETTE
CONTROL

ON/OFF/TIMER
POWER SWITCH

ON/OFF POWER SWITCH
(MASTER CONTROL UNIT)

GRASPING HAND

POWER LIGHT

DIRECTIONAL
STICK

DIRECT MIC HOOKUP

ROBOT SOUNDS

REMOTE
MICROPHONE

BATTERY CHARGER
LIGHT

CASSETTE UNIT

MASTER
CONTROL UNIT

ONBOARD
SPEAKER

HOME BASE
ARROW

The Omnibot and its control unit.

Setting Up and Using the Omnibot

Before you can make the Omnibot move, you have to charge the built-in sealed lead-acid battery, which takes at least 12 hours when fully discharged. A battery charger is included, but you also need six AA-size batteries that are not included—two to power the robot's digital clock and four more for the master control unit.

The Omnibot has three modes of operation: tape, program and remote control. The tape mode is for using the cassette deck to play prerecorded audio tapes or to record music or speech onto blank tapes. The quality of the sound reproduction is surprisingly good, considering that the primary purpose of the tape deck is to record programming instructions (as we will see when we take a look at the program mode).

The remote-control mode is for steering the robot forward, backward, right and left using the master control unit's joystick. When you speak into the control unit's microphone, your voice is projected from the robot's built-in speaker. In this mode, you can also start and stop the cassette tape and select two different special sound effects.

In program mode, all steering commands and sound effects are recorded on the cassette tape for playback later. Also, anything you say into the remote microphone is recorded on the tape. The effect, on playback, is remarkable. It is literally child's play to program the Omnibot to go from the kitchen into the living room, carrying some snacks on its serving tray, and stopping in front of the couch to ask if anyone would like something to eat. The Omnibot is supposed to be a toy, but this feat is well beyond the capabilities of some personal robots that are *not* sold as toys. The Omnibot can also be programmed to play a cassette tape—containing either music and speech or motion commands—at a preselected time on any given day of the week.

Suggestions for games to play with the Omnibot can be found in the back of the instruction manual. Some examples are racing the robot through an obstacle course, programming the robot to leave a message for someone in your family, and recording half of a dialogue on tape so that on playback the robot appears to be having a conversation with you.

The Omnibot has a suggested retail price of $250—on the high side for a toy, but very reasonable for a small personal robot.

16

Personal Robot Accessories

Virtual Devices' Menos 1

Menos 1 is a hardware and software add-on package for Heath's Hero 1 personal robot. The package consists of an expansion board that replaces the microprocessor chip in Hero's "brain," an RS-232 or remote radio data link, and a floppy disk with software that runs on a personal computer. The software is available in two versions, one for MS-DOS and one for CP/M-compatible systems.

With this package, you can operate the Hero 1 from a personal computer. You can also monitor the robot's operation on the computer's CRT screen, using a sophisticated split-window feature. The software consists of four main modules which provide remote operation, simulation, editing and file management.

- Remote Operation: This module lets you work with your personal computer keyboard as if it were the Hero 1 keypad. The computer's CRT shows you exactly what you would see if you were looking at the robot's LED display panel. You can also download Hero 1 programs directly from your computer.

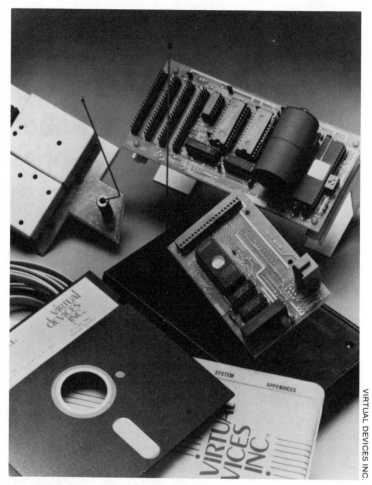

Virtual Devices' Menos 1 system.

• Simulation: You can simulate the running of a Hero 1 program on your computer with this module, using the CRT screen to keep track of changes in motor positions, sensor readings and memory contents. You can run the simulation at high speed, low speed, or one step at a time.

• Editing: This module consists of program generation and editing aids.

• File Management: All your file operations, such as saving new programs and reading previously stored files, are handled by this module.

Menos 1 is a useful and desirable add-on for Hero 1 owners because it takes the tedium and drudgery out of programming the robot. The performance of the simulation module, in particular, has been compared very favorably with that of highly sophisticated simulation systems used in military and aerospace applications. The price of the Menos 1 package is $595 for the RS-232 version, and $795 for the remote radio version.

Arctec Systems
Arctec Systems manufactures several products designed to make it easier for the hobbyist to program a Heath Hero 1 robot. The three Arctec products we will discuss here are the Hero Memcom board, the Voice Command System, and the Apple-Hero Communicator.

The Hero Memcom Board
The Memcom (*Mem*ory and *Com*munications) board increases the total amount of memory available in the Hero 1 to 32K bytes, and also adds a serial communications port. The main purpose of the board is to eliminate the need to hand-code programs for the robot. Instead, you can use a personal computer to write your programs, then download them to the robot. The Memcom board, which is a prerequisite if you want to add the Voice Command System, is priced at $345.

The Voice Command System
With this Hero 1 add-on, you can forget about typing in hexadecimal codes, and use spoken, English-language commands instead. When the Vorec (*Voice Rec*ognition) board is installed, you get following menu—*spoken by the Hero 1 robot*, rather than displayed on a computer monitor:

> "Push key 1   to train,
>   2   for deferred execution
>   3   for immediate execution
>   4   to retrain
>   5   to abort."

Voice Training Mode: The first step is training the system to recognize your voice. To do this, you push key 1 on the robot's keypad to enter training mode, then stand about three feet away from the robot and speak each command word in turn. The system makes it easy for you by using the robot's voice synthesis capability to prompt you for each command. Thus, the robot will start by saying "Say 'forward.'" After you repeat the word "forward," the robot will ask you to "Say 'back.'"

You will be asked to repeat each word in the system's command vocabulary enough times for the system to be able to determine the frequency range of your voice and to store a typical sound pattern corresponding to each command word. There are 32 individual command words in the vocabulary, but many more distinct commands can be created by stringing words together in different combinations.

When the training session is complete, the system returns to the main menu.

Deferred Execution Mode: You enter this mode by pressing key 2 on the robot's keypad. The command words you can use in this mode are:

> Forward
> Back
> Right
> Left
> Stop
> Go
> Menu
> and the digits "Zero" through "Nine"

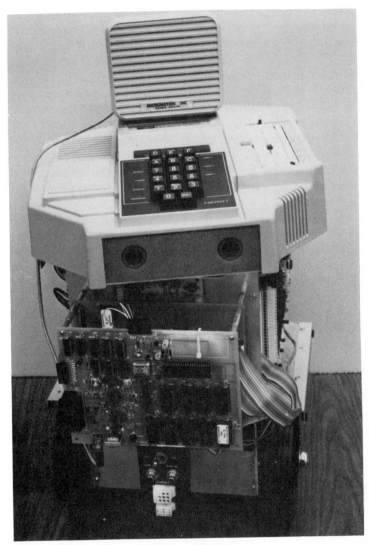

Arctec Systems' voice command system.

After issuing the commands "Forward" or "Back," you must specify a two-digit number from "01" to "99" corresponding to the number of "robot-lengths" you want the robot to move. Similarly, the commands "Right" and "Left" must be followed by "01" through "99"—the number of degrees that you want the robot to turn relative to the present heading.

When you are done programming, you use the command "Stop" to signify the end of the program. To run the program, you just say "Go."

The voice recognition system repeats each word it recognizes to make sure it is interpreting your commands correctly. You can then say "Roger" to acknowledge a correct interpretation or "No" or "Negative" if the interpretation was wrong.

Immediate Execution Mode: In this mode, selected by pressing key 3 from the main menu, you can program arm movements in addition to the forward, back, right and left body movements available in deferred mode. The commands in immediate execution mode are:

Command	Option	Description
Body	Forward	Move body
	Back	
	Right	Rotate body
	Left	
Head	Right	Rotate head
	Left	
Shoulder	Up	Rotate shoulder
	Down	
Wrist	Right	Rotate wrist
	Left	
	Up	
	Down	
Grip		Close gripper
Release		Open gripper

Retrain Mode: If the Voice Command System is consistently having trouble recognizing certain words, you can press key 4 from the main menu to retrain a few words without having to go through the entire command word list.

The instruction manual includes some suggestions for improving the reliability of the system, which the manufacturer says is 98% accurate under ideal conditions. It's important to speak clearly without mumbling and without any extraneous lip and breathing noises. It also helps to try different words (such as "Yeah" instead of "Roger") if particular words consistently cause problems.

The Vorec board is priced at $595.

The Apple-Hero Communicator

The Apple-Hero communicator board is a general-purpose I/O and memory expansion board. It is particularly useful for allowing high-speed data transfer between an Apple computer and a Hero 1 robot equipped with the Memcom board. Data transfer can take place at much higher speeds than usual because the board uses parallel communications rather than serial.

The Apple-Hero Communicator board is priced at $159.

Personal Robot

Robot	Voice Synthesis	Functioning Arm	Preprogrammed Software Available
Topo	Yes	No	No
Hero 1	Optional	Optional	No
RB5X	Optional	Optional	Yes
Hubot	Yes	No	Yes
Hero Jr.	Yes	No	Yes
Turtle Tot	Optional	No	Yes
FRED	Yes	No	No
Dingbot	No	No	No
Verbot	No	No	No
Omnibot	No	No	No

Comparison Table

Can Be Used Without Separate Microcomputer	Base Price (Including Voice, If Available)	Other Options Available
No	$1,595	Snack tray
Yes	2,200 (incl. arm)	Robotics course
Yes	2,690	RCL software
Yes	3,500	Small printer
Yes	1,000	Radio control
No	500	None
Yes	500	Battery charger
Yes	10	None
Yes	60	None
Yes	250	None

Personal Robot Industry Resources

Manufacturers

Androbot Inc.
101 E. Dagget Dr.
San Jose, CA 95134
Topo, FRED and BOB

Android Industries Inc.
P.O. Box 1240
Morgan Hill, CA 95037
Show robots

Arctec Systems
9104 Red Branch Rd.
Columbia, MD 21045
Personal robot accessories

Bingel Robotics
P.O. Box 14655
Gainesville, FL 32604
Educational robots

Colne Robotics Inc.
P.O. Box 23416
Fort Lauderdale, FL 33307
Educational robots

ComRo Inc.
126 E. 64th St.
New York, NY 10021
ComRo Tot personal robot

Economatics (Education) Ltd.
4 Orgreave Cres.
Handsworth, Sheffield, England
BBC Buggy personal robot

Excalibur Technologies Corp.
800 Rio Grande Blvd. NW
Albuquerque, NM 87104
Software for personal robots

Feedback Inc.
620 Springfield Ave.
Berkeley Heights, NJ 07922
Educational robots

Flexible Systems
219 Liverpool St.
Hobart, Tasmania, Australia
Turtle Tot personal robot

Heath Company
Benton Harbor, MI 49022
Hero 1 and Hero Jr.

Hobby Robot Co. Inc.
P.O. Box 887
Hazlehurst, GA 31539
Hobbyist robots

Hubotics Inc.
5375 Avenida Encinas
Carlsbad, CA 92008
Hubot personal robot

Iowa Precision Robotics
908 10th St.
Milford, IA 51351
Marvin personal robot

Microbot Inc.
453-H Ravendale Dr.
Mountain View, CA 94043
Educational robot arms

Nationwide Robots
4141 Woerner Ave.
Levittown, PA 19057
Show robots

OWI Inc.
1938-A Del Amo Blvd.
Torrance, CA 90501
Movit toy robot kits

Personal Robotics Corp.
469 Waskow Dr.
San Jose, CA 95123
Hobbyist robots

RB Robot Corp.
18301 W. 10th Ave., #310
Golden, CO 80401
RB5X personal robot

Rhino Robots Inc.
2505 S. Neil St.
Champaign, IL 61820
Educational robots

Robot Company
2189 NW 53rd St.
Fort Lauderdale, FL 33309
Show robots

Robot Factory
P.O. Box 112
Cascade, CO 80809
Show robots

Robot Repair
816 21st St.
Sacramento, CA 95814
Hobbyist robots

Robot Shack
P.O. Box 582
El Toro, CA 92630
Hobbyist robots

RobotEx
111 E. Alton Ave.
Santa Ana, CA 92707
Hobbyist robots

Robotronix Inc.
P.O. Box 1125
Los Alamos, NM 87544
Personal robotics software

Spectron Instrument Corp.
1342 W. Cedar Ave.
Denver, CO 80223
Hobbyist robots

Tomy Toys
P.O. Box 6252
Carson, CA 90749
Dingot, Verbot & Omnibot

Virtual Devices Inc.
P.O. Box 30440
Bethesda, MD 20814
Personal robot accessories

Retailers and Distributors

Amarobot
1780 Shattuck Ave.
Berkeley, CA 94709
Personal robot store

Harvard Associates Inc.
260 Beacon St.
Somerville, MA 02143
U.S. Distributor for Turtle Tot

Interface Technology Inc.
P.O. Box 3040
Laurel, MD 20708
Hobbyist robot components

New Tech Promotions
2265 Westwood Blvd., Suite 248
Los Angeles, CA 90024
Personal robot distributor

Rio Grande Robotics
1595 W. Picacho #28
Las Cruces, NM 88005
Personal robot distributor

Robotland
4251 N. Federal Hwy.
Boca Raton, FL 33431
Personal robot store

Robotorium
252 Mott St.
New York, NY 10012
Toy robot store

Periodicals

Personal Robotics News
P.O. Box 10058
Berkeley, CA 94709
Industry newsletter

Robot Insider
11 E. Adams St., Suite 1400
Chicago, IL 60603
Industrial robotics newsletter

Robotics Age
174 Concord St.
Peterborough, NH 03458
Robotics magazine

Robotics Today
P.O. Box 930
Dearborn, MI 48121
Industrial robotics magazine

Robotics World
6255 Barfield Rd.
Atlanta, GA 30328
Industrial robotics magazine

Smart Machines
P.O. Box 459
Sharon, MA 02067
Robotics newsletter

Associations

Homebrew Robotics Club
91 Roosevelt Circle
Palo Alto, CA 94306
Robotics hobbyist club

ROBIG
3205 Sydenham St.
Fairfax, VA 22031
Personal robotics interest group

Robotic Industries Association
P.O. Box 1366
Dearborn, MI 48121
Robotics industry association

Robotics Society of America
200 California Ave., Suite 215
Palo Alto, CA 94306
Personal robot association

Glossary

analog-to-digital converter A circuit for converting electrical voltages representing physical quantities into numbers so that a computer can process them.

applications software Software for making a robot perform a specific function or application.

artificial intelligence The ability of a machine to perform functions that would require intelligence if they were performed by a human being.

assembly language A symbolic representation of machine code.

automaton Mechanical device that can perform certain actions without human intervention.

Basic A relatively simple computer programming language. See also *Tiny Basic.*

binary code Sequences of 0s and 1s used to represent numbers and characters.

bits *bi*nary dig*its*, i.e., a sequence of digits set to various combinations of 0s and 1s.

byte The basic unit of measurement of computer memory. Usually equal to eight bits.

chip Small piece of semiconducting material on which all the circuits of an entire computer can be built.

compliance The ability of a robot's gripper to "give" slightly so that it can pick up an object that doesn't precisely fit the shape of the gripper.

computer-aided instruction The use of a computer to present educational material and to test how well a student has absorbed the material.

CP/M Control Program for Microcomputers, a widely used operating system for microcomputers.

download Transmit programs developed on a computer to another computer or a robot.

expert systems Computer systems consisting of large quantities of stored information capable of applying that information to solving problems (e.g., diagnosis of rare diseases).

feedback Information about the results of past actions that can be used in deciding what to do next.

frame One complete "still" picture extracted from a changing scene.

gripper A robotic hand, often consisting of only two or three fingers that close together.

hardware The physical components that make up a computer or robot (see also **software**).

hexadecimal code A system of representing numbers inside a computer using 16 digits rather than just the 10 that we normally use.

image processing The technology of capturing a scene with a camera and using a computer to interpret that scene.

infrared sensor A sensor for measuring infrared light, which is invisible to human beings.

LED Light-emitting diode, a semiconductor that emits light when current flows through it. Often used as an indicator light.

Logo A computer language designed for introducing children to computer programming.

machine language The binary codes that all programming instructions must eventually be translated into before computers can execute them.

memory The components and circuitry that enable a computer to store and retrieve information.

microcomputer A microprocessor plus the memory, input/output interfaces and power supply necessary for the device to function as a small computer.

microprocessor The central processing unit portion of a complete computer system, built on a single chip.

MS-DOS A popular operating system for the IBM PC and compatible computers.

operating system Software that manages a computer system's hardware and software resources and acts as an interface between a computer system and the user.

parallel port A means of receiving or transmitting data in which all the bits in a byte can be processed simultaneously (see also *serial port*).

personal robot A robot designed to be owned and used by an individual.

phoneme A sound representing a portion of a word.

robot A machine that can both sense and manipulate its environment.

robotic Of, or pertaining to, a robot.

robotics The science of designing, building and programming robots.

RS-232 An internationally recognized standard for receiving or transmitting data via a serial port (see also *serial port*).

semiconductor A device that can act either as an electrical conductor or as an insulator, and can therefore be used as a switching device within a computer.

sensor A device for providing information about the environment to a robot.

serial port A means of receiving or transmitting data in which all the bits in a byte are processed one at a time (serially). See also *parallel port*.

serial communications card A circuit board inserted into a computer to enable it to input and output data serially.

show robot A radio-controlled machine that looks and acts like a robot, but is in fact controlled by a human operator.

sleep switch A switch for deactivating a robot without losing the contents of the robot's memory.

software The programming instructions that tell a computer or robot what to do (see also **hardware**).

sonar A sensing system that measures distance using an ultrasonic transmitter and receiver.

speaker-dependent The type of voice-recognition system that recognizes only one voice.

speaker-independent The type of voice-recognition system that will recognize the voices of many different people.

speech recognition The ability of a robot to understand human speech.

speech synthesis The ability of a robot to generate human-sounding speech.

storage Same as *memory*.

teaching pendant A control box used to walk a robot through a sequence of motions which it can then repeat over and over again.

text-to-speech converter A system for converting words stored in memory into synthesized speech.

Tiny Basic A highly compact programming language useful for small microcomputers with very limited storage capacity.

transducer A device for measuring physical quantities and converting that measurement into electrical signals.

ultrasonic sensor A sensor for measuring high-frequency sound, which is beyond our hearing range.

voice training The process of exposing a voice recognition system to the particular sound of your voice so that it will be able to recognize your speech.

Bibliography

Asimov, Isaac. *The Robot Collection.* Doubleday & Co., 1983. 567 pages. Fiction. Contains Asimov's first two celebrated robot novels, *The Caves of Steel* and *The Naked Sun*, and every short story Asimov has ever written about robots.

Asimov, Isaac. *The Robots of Dawn.* Doubleday & Co., 1983. 430 pages. Fiction. The third in Asimov's series of novels about a detective and his robot assistant investigating a murder.

Compton, Michael M. *Understanding Robots.* Alfred Publishing Co., 1983. 47 pages. A brief look at commercially available hobby robots, including a list of references and sources for further information.

Davies, Owen, Editor. *The Omni Book of Computers and Robots.* Kensington Publishing Corp. 413 pages. Articles on computers and robots, excerpted from *Omni* magazine.

Geduld, Harry M., and Gottesman, Ronald. *Robots, Robots, Robots.* 246 pages. A collection of stories, plays, essays and pictures which describe the fascinating history of the "mechanical man" concept.

Gloess, Paul. *Understanding Artificial Intelligence.* Alfred Publishing Co., 1981. 47 pages. A brief, interesting introduction to the fundamentals of artificial intelligence and some of its major applications.

Graham, Neil. *Artificial Intelligence.* TAB Books, 1979. 251 pages. An introductory level book on the science of artificial intelligence and its applications.

Gupton, James. *Microcomputers for External Control Devices.* Dilithium Press, 1980. 279 pages. An introductory level text on the use of microprocessors for external control of mechanical devices.

Heiserman, David L. *Projects in Machine Intelligence for Your Home Computer.* TAB Books, 1982. 337 pages. Explanations and demonstrations (using Basic language programs) of the behavior of intelligent machines.

Heiserman, David L. *How to Design and Build Your Own Custom Robot.* TAB Books, 1981. 462 pages. Provides the information you need to design and build your own robot.

Heiserman, David L. *How to Build Your Own Self-Programming Robot.* TAB Books, 1979. 237 pages. Describes an advanced self-programming robot called Rodney, with step-by-step instructions on building and programming a Rodney of your own.

Heiserman, David L. *Robot Intelligence With Experiments.* TAB Books, 1981. 322 pages. Describes the author's development of approaches to the computer simulation of robot intelligence.

Heiserman, David L. *Build Your Own Working Robot.* TAB Books, 1976, 234 pages. Tells you how to build Buster, an independent mobile robot.

Helmers, Carl T., Editor. *Robotics Age: In the Beginning.* Hayden Book Co., 1983. 241 pages. A collection of the best technical articles appearing in *Robotics Age* magazine from 1979 to 1981.

Holland, John M. *Basic Robotic Concepts.* Howard Sams, 1983. 270 pages. An in-depth study of robotics components, including motors, manipulators, mobility, and vision systems.

Krutch, John. *Experiments in Artificial Intelligence for Small Computers.* Howard Sams, 1981. 110 pages. This book uses programs written in Basic to demonstrate artificial intelligence applications.

Lonergan, Tom, and Fredericks, Carl. *The VOR (Volitionally Operant Robot).* Hayden Book Co., 1983. 120 pages. A discussion of the philosophy of robotic design.

Loofbourrow, Tod. *How to Build a Computer-Controlled Robot.* Hayden Book Co. 132 pages. Provides detailed instructions for building a robot controlled by a KIM-1 microprocessor.

Malone, Robert. *The Robot Book.* Push Pin Press, 1978. 159 pages. An illustrated book exploring the role of robots as early mechanical toys, as heros and villains in books and films, as industrial workers and as the creations of individual inventors.

McCorduck, Pamela. *Machines Who Think.* W.H. Freeman & Co. 375 pages. A history of the highly technical subject of artificial intelligence, written in an entertaining style.

Robillard, Mark J. *Hero 1: Advanced Programming and Interfacing.* Howard Sams, 1983. 233 pages. Specialized information on programming the Heath Hero 1 and adding hardware enhancements.

Safford, Edward L., Jr. *Handbook of Advanced Robotics.* TAB Books, 1981. 468 pages. A wide range of topics of interest to the robotics hobbyist, including suggestions on design considerations.

Safford, Edward L., Jr. *The Complete Handbook of Robotics.* TAB Books, 1980. 358 pages. Looks at how hobby robots are put together, and how they work.

Schmidt, Neil, and Farwell, Robert. *Understanding Electronic Control of Automation Systems.* Texas Instruments Learning Center, 1983. 280 pages. An introduction to the basic concepts of how electronics are used to control machines.

Warring, R.H. *Robots and Robotology.* TAB Books, 1984. 128 pages. Overview of robot technology with particular emphasis given to first- and second-generation industrial robots.

Weinstein, Martin Bradley. *Android Design: Practical Approaches for Robot Builders.* Hayden Book Co., 1981. 248 pages. Emphasis on sophisticated robot design and construction using components available to the hobbyist.

Index

About the Author

Mike Higgins is the editor of *Personal Robotics News,* a monthly newsletter for manufacturers and vendors of personal robots. Before entering the publishing field he spent ten years in the computer industry, holding various positions in programming, sales and marketing.

Born in Birmingham, England, the author graduated from the University of Essex with a degree in computer science before moving to the United States. He now lives with his wife, Marlene, in Berkeley, California.